Manual of Space Security

Attention Authors

'We Convert Fighters into Writers'. That's the motto of Manas Publications since more than four decades. Our message to the world is that India can win the battle not only by the bullet but also by the pen. There is no dearth of intellectuals in our country, but their knowledge is confined to them only. No sincere effort has been made by any other publisher to channelise their knowledge in national interest. That is why the talent of a large number of intellectuals remains latent. An author transforms his knowledge into a manuscript, and it is our job to spread his/her knowledge by converting it into a beautiful book. For over four decades, Manas has been motivating the intellectuals by publishing their manuscripts and would like to publish your manuscript too. If you or your colleagues have any manuscript ready for publication or are working on it, kindly contact us with detailed synopsis and list of contents. We will transform your work into a valuable publication. If you wish, we can also suggest you the suitable subjects for writing the books related to your area of expertise. Manas is known for its publishing quality and commitments. We take utmost care in production and give wide publicity to the book as well as its author through our National and International distribution network.

MANUAL OF SPACE SECURITY

David Wright
Laura Grego
Lisbeth Gronlund

Manas Publications
(We Convert Fighters into Writers)
New Delhi-110002 (INDIA)

MANAS PUBLICATIONS
(We Convert Fighters into Writers)
(Publishers, Distributors, Importers & Exporters)
4402/5-A, Ansari Road (Opp HDFC Bank)
Darya Ganj, New Delhi-110 002 (India)
Off. ✆ 011-23260783, 23265523, Res. ✆ 011-23842660
Fax: 011-23272766
E-mail: manaspublications@gmail.com
manaspub6@gmail.com
Website: www.manaspublications.in

ISBN 978-81-7049-452-2
₹ 995/-

Designed at
Manas Publications

Printed in India at
RK Offsets, Delhi
and published by Mrs Suman Lata for
Manas Publications, 4402/5-A, Ansari Road,
(Opp HDFC Bank),
Darya Ganj, New Delhi-110 002 (INDIA)

Contents

Foreword

Over the past decade or so, societies around the world have relied increasingly on satellites for vital communication services, environmental monitoring, navigation, weather prediction, and scientific research. This largely beneficial trend is expected to intensify: more countries are developing satellite technology and using the services derived from it.

The same technologies have made possible the development of military capabilities in space that go far beyond those employed during the Cold War for intelligence gathering early warning. Some in the United States see space as a critical enabler for bringing decisive military force to bear anywhere on Earth with little or no warning. This rapid strike capability is a central element of the post-9/11 national security strategy, which seeks not only to deter or defeat any potential aggressor but also to prevent the acquisition of threatening capabilities by hostile states or terrorist groups. Protecting and enhancing U.S. military capability in space is emerging as an important focus of military planning. Recent official documents have proposed, for example, various anti-satellite and space-based weapons to protect and augment U.S. capabilities in space.

These new missions are controversial in the view of close U.S. allies and are likely to be contested by others if pursued. Serious public discussion of military space plans has not yet occurred in the United States, though important questions of policy, planning and budgeting loom: What missions are best carried out from space? What are the likely costs and available alternatives to various space weapons proposals? How susceptible are satellites to interference? How easily can they be disabled or destroyed? What measures can be taken to reduce their vulnerability?

The answers to these questions depend on physical laws and technical facts that are not widely understood outside of a rather narrow slice of the science and engineering community. The paper that follows makes accessible to a general audience the necessary facts upon which an informed evaluation of space policy choices can take place. The authors, physicists David Wright, Laura Grego, and Lisbeth Gronlund, describe the mechanics of satellite orbits and explain why certain operations are suited to particular orbits. They discuss the requirements for launching satellites into space and maneuvering them once in space. They consider the consequences of the space environment for basing certain military missions there. Finally, they describe the elements of a satellite system and assess the vulnerability of these components to various types of interference or destruction. They also include an analysis of technical measures for reducing satellite vulnerability.

The paper makes no attempt to provide policy recommendations. Although the authors' views on space weapons and missile defense are well known to those who follow these issues, they are not asserted here. Instead, the intent is to provide a neutral reference. Those engaged in the policy process, no matter what their views, should find this work useful.

This paper is part of the American Academy's "Reconsidering the Rules of Space" project. The study examines the implications of U.S. policy in space from a variety of perspectives, and considers the international rules and principles needed for protecting a long-term balance of commercial, military, and scientific activities in space. The project is producing a series of papers, intended to help inform public discussion of legitimate uses of space, and induce a further examination of U.S. official plans and policies in space. Forthcoming papers will consider the interaction of military, scientific, and commercial activities in space; Chinese and Russian perspectives on U.S. space plans; and the possible elements of a more comprehensive space security system.

The authors presented parts of the paper at a workshop convened by the American Academy and its Committee on International Security Studies in December 2003. Participating were Bruce Blair, Steve Fetter, Nancy Gallagher, Richard Garwin, Subrata Ghoshroy, Joan Johnson-Freese, Carl Kaysen, George Lewis, Martin Malin, Jonathan McDowell, Norman Neureiter, Pavel Podvig, Theodore Postol, John Rhinelander, John Steinbruner, Eugene Skolnikoff, Larry S. Walker, and Hui Zhang. We thank the participants for their insights at the workshop.

We also thank Richard Garwin, George Lewis, and two anonymous reviewers for their comments on the paper. We join the authors in thanking Steve Fetter, Richard Garwin, George Lewis, Pavel Podvig, and Wang Ting for providing additional comments. And we join them in acknowledging Helen Curry, Phyllis Bendell, Andrea Grego, and Anita Spiess for their efforts in producing this report. We are grateful to the authors for their work on the paper and to the Union of Concerned Scientists for providing the authors with the time and space to carry out their work.

The Rules of Space project at the American Academy is supported by a generous grant from the Carnegie Corporation of New York. We thank the Carnegie Corporation for its support and Patricia Nicholas for her assistance.

John Steinbruner
University of Maryland

Carl Kaysen
Massachusetts Institute of Technology

Martin Malin
American Academy of Arts and Sciences

Section - 1

Introduction

In the nearly fifty years since the Soviet Union launched Sputnik, there has been a steady growth in the number of countries that have launched satellites into orbit. Growing even faster is the number of countries that have deployed satellites launched by others. Currently, satellites serve a multitude of civilian and military functions, from facilitating communications and weather forecasting to providing highly accurate navigational information, and many nations envision making future investments in satellites for such uses.

In the U.S. military there is also a growing interest in basing weapons in space as well as in developing means to attack the satellites of other nations and to protect U.S. satellites from attack. While space has long been home to military systems for observation, communication, and navigation, these new missions would be a departure from long-held norms. There are currently no known weapons stationed in space that are explicitly designed to apply force. Nor are there any known deployed systems designed explicitly to destroy satellites, either from the ground or from space.

This shift in U.S. military thinking is evident from planning documents released in recent years that envision a restructuring of military commands and the development and deployment of anti-satellite weapons and spacebased weapons.[1] These new systems are meant to fulfill four general missions:

1. See, for example, Report of the Commission to Assess United States National Security, Space Management and Organization, January 11, 2001, http://www.fas.org/spp/military/commission/ report.htm, accessed February 8, 2005, and Air Force Space Command, "Strategic Master Plan: FY06 and Beyond," October 1, 2003, http://www.peterson.af.mil/hqafspc/news/ images/ FY06%20Beyond%20Strategic%20Master%20Plan.pdf, accessed February 8, 2005.

- defending U.S. satellites and ensuring U.S. freedom to operate in space
- denying adversaries the ability to use space assets
- intercepting ballistic missiles using space-based interceptors
- attacking targets on the ground or in the air using space-based weapons

The first two missions reflect the military importance of current U.S. space-based systems. This utility has led to a desire to protect these systems and to deny similar capabilities to potential adversaries. The third mission, an ongoing interest of many missile defense proponents, is leading toward the deployment of prototype weapons as part of a space-based "test bed." The fourth mission, which has attracted considerable public attention and concern, currently appears to be of less interest to the U.S. military than the other missions.

U.S. interest in new types of weapons has spawned an emerging international debate. Key topics include whether the deployment of space-based weapons and anti-satellite weapons is inevitable, what military utility such weapons would have, how their deployment would affect the security of the owner nation and the wider international community, whether their deployment and use would interfere with other military and civilian uses of space, and what normative and legal constraints on the use of space could be agreed upon and enforced.

Addressing these issues requires assessing a wide range of political, diplomatic, military, and technical issues. This report is limited to a discussion and analysis of the technical and military issues and focuses on a number of key questions: What capabilities could anti-satellite weapons and weapons in space realistically provide? Would these capabilities be unique? How do they compare with alternatives? What would they cost? What options would be available to nations seeking to counter these capabilities? The answers depend on technical realities that must be considered in any policy analysis of space weapons and anti-satellite weapons. Unless debate about these issues is grounded in an accurate understanding of the technical facts underlying space operations, the discussion and policy prescriptions will be irrelevant or, worse, counterproductive.

In evaluating proposed military systems, it is important to distinguish between constraints imposed by financial cost, by technology, and by physics. The cost of operating in space is often high relative to the cost of operating in the air or on the ground. While cost will be important in considering development and deployment, it may not be decisive if the system could provide a unique capability that is deemed important. Available technology places important limits on what systems are currently feasible for a given country, but those limits can change over time and do not represent fundamental limitations. The space-based laser, for example, has so far achieved power levels well below what is required for a usable weapon, but there do not appear to be fundamental limits to increasing its power over time. Physics, on the other hand, places fundamental limits on space operations that will not change with time. An example of a fundamental limit posed by physics is the fact that satellites in low orbits cannot remain stationary over a given location on Earth, so multiple satellites are required to ensure that one is always near that location.

This report provides information on a range of technical issues related to space systems that are important for anyone involved in the debate over space security to understand.[2] It discusses

2. Other books and articles that address technical issues related to space security include Ashton B. Carter, "Satellites and Anti-Satellites: The Limits of the Possible," International Security (Spring 1986): 46-98; Richard L. Garwin, "Space Technology: Myth and Promise," 1988, http://www.fas.org/rlg/myths-of-space.htm, accessed February 8, 2005; Michael J. Muolo et al., Space Handbook, Volume 1: A War Fighter's Guide to Space, (Maxwell Air Force Base, AL: Air University Press, December 1993), http://www.au.af.mil/au/awc/awcgate/au- 18/au180001.htm, accessed February 8, 2005; Michael J. Muolo et al., Space Handbook, Volume 2: An Analyst's Guide, (Maxwell Air Force Base, AL: Air University Press, December 1993); Office of Technology Assessment, Anti-Satellite Weapons, Countermeasures, and Arms Control (Washington, DC: Government Printing Office, 1985); Philip E. Nielsen, Effects of Directed Energy Weapons, (National Defense University, 1994), http://www.ndu.edu/ctnsp/directed_ energy.htm, accessed December 23, 2004; James Oberg, Space Power Theory, 1999, http://www. jamesoberg.com/spt.html, accessed February 8, 2005; Bob Preston et al., Space Weapons Earth Wars (Arlington, VA: RAND Project Air Force, 2002) http://www.rand.org/publications/MR/ MR1209/, accessed February 8, 2005; Report of the American Physical Society Study Group on Boost-Phase Intercept Systems for National Missile Defense, July 2003, http://www.aps.org/public_ affairs/ popa/reports/nmd03.html, accessed

issues of cost and technology, where appropriate, and attempts to separate these from the fundamental issues of physics. It is written for a lay audience but includes appendices that give more detailed technical information for specialists. The report is also intended to familiarize readers with the important technical terminology and concepts related to satellites and operating in space. For example, the behavior of objects traveling at very high speeds in space is much different than the behavior of objects in motion on the ground or in the atmosphere and is largely outside day-to-day human experience. As a result, most people have not developed intuition about the behavior of satellites, so that attempting to apply lessons from common experience can lead to mistakes and misconceptions. In addition, the report shows that a few basic laws of physics have important implications for the way satellites, space-based weapons, and antisatellite weapons can be designed and operated. It explains these underlying physical principles and discusses their implications.

The report addresses technical issues that are relevant to space policy, but does not address policy issues per se or make policy recommendations. The report is not intended to be comprehensive; the omission of a topic should not be construed to mean the topic is not important.

Sections 2 and 3 lay out the main points of the paper. Section 2 presents the report's findings and conclusions that have implications for space policy and directs the reader to the sections from which these conclusions are derived. Section 3 summarizes the main technical points made in the subsequent sections.

Sections 4 through 8 discuss basic concepts and implications of orbital dynamics; Section 4 covers the basics of satellite orbits; Section 5 inventories the types of orbits and the criteria for choosing a particular orbit; Section 6 discusses the physics of maneuvering in space; Section 7 assesses the implications of

---

February 8, 2005; Air University Space Primer, August 2003, http://space.au.af.mil/primer/, accessed February 8, 2005; Federation of American Scientists' Panel on Weapons in Space, Ensuring America's Space Security, September 2004, http://www.fas.org/main/content.jsp?formAction=297&contentId=311, accessed February 8, 2005; Bruce M. DeBlois et al., "Space Weapons: Crossing the U.S. Rubicon," International Security (Fall 2004): 1–34.

this maneuvering for satellite mass; and Section 8 discusses the physics and technology of launching mass into space and placing satellites in orbit.

Section 9 examines the implications of these technical assessments for several specific space-based systems—in particular, space-based constellations of ground-attack weapons, space-based missile defense interceptors, and the military space plane. The final three sections discuss interference with satellite operations. Section 10 identifies and discusses the various components that constitute a satellite system, because different methods of interference target different components. Section 11 gives an overview of many of the possible means to interfere with a satellite system. Section 12 looks in more detail at three particular topics related to interference: space-basing of anti-satellite (ASAT) weapons, a simple ASAT that would place debris in the path of a satellite, and ways to mitigate satellite system vulnerability.

Section - 2

Policy-Relevant Implications

This report discusses how the laws of physics apply to operations in space and to interference with such operations. The essential technical facts that emerge from that discussion are summarized in Section 3. This section lays out some of the policy-relevant implications of these technical facts for the operation and utility of various systems, including space-based weapons, as well as antisatellite weapons (ASATs), both ground based and space based.

Space-based weapons can be designed to destroy their targets in one of several ways: with direct impact,[1] an explosive warhead, or a laser. Spacebased lasers intended to damage parts of satellites other than the sensors are not discussed in this report because the technology for these weapons will not be available in the foreseeable future.

The first five points below concern the types of missions for which space is or is not well suited. Points 6 through 9 address anti-satellite weapons, the ASAT capabilities of systems designed for other purposes, and the vulnerability of satellites. Point 10 concerns the overall military utility of space-based weapons. Each point notes the section or sections of the report where the issue is discussed.

1. Space basing is uniquely well suited to a wide range of civilian and military applications. (Sections 5, 6, 7)

Space offers several features not available from the ground or air. Satellitebased sensors can see much larger areas of the Earth

1. Weapons that destroy by direct impact are called kinetic kill weapons. The kinetic energy of the fast-moving weapon and/or target provides the energy to destroy the target.

than sensors closer to the Earth can see. This allows large-scale simultaneous observation of the Earth's surface and atmosphere, and communication between and simultaneous broadcast to large parts of the earth.

Because the atmosphere blocks transmission of many types of electromagnetic waves, some kinds of astronomical observations can only be made from space.

Moreover, space is much better suited to some types of operations than to others. Electromagnetic signals (light and radio waves) can be transmitted over large distances almost instantaneously and with very little energy cost. Space therefore favors activities that entail sending and receiving electromagnetic signals over activities that involve transporting large amounts of mass from the Earth into space or that involve significant maneuvering in space, which can require a large mass of propellant.

As a result, the applications for which space basing is uniquely well suited include:

- large-scale environmental monitoring of, for example, atmospheric behavior, climate change, and deforestation
- large-scale weather monitoring for weather forecasting
- astronomy
- global communication, broadcast, and data transfer
- highly accurate navigation and position determination
- reconnaissance on a global or large-scale basis
- detection on a global basis of missile launches, to provide early warning of attacks and information about the missile testing programs of nations

Some of these tasks could, as discussed later, be accomplished by ground- and air-based alternatives, if urgently needed, albeit on a regional rather than global basis.

2. Space basing is poorly suited to the mission of attacking ground targets using kinetic energy or explosive weapons. This mission can be done as well or better from the ground, and acquiring a prompt attack capability from space would be much more costly. (Section 9)

Space-based ground attack weapons could offer global reach, obviating the need to forward-base weapons, providing prompt attack capability, and shortening any warning of an attack. However, space-based weapons fare poorly when compared with other long-range means of attacking ground targets. In particular, intercontinental-range ballistic missiles can provide the same prompt, global reach, but are less expensive and more reliable than spacebased weapons.

Cost

Deploying a space-based system would be tens of times more expensive than deploying a comparable system using ballistic missiles. This is a consequence of the fact that any satellite system with prompt global coverage would require numerous satellites to ensure that at least one is in the right place at all times. The exact number of satellites will depend on the altitude of the orbit and the reach of each weapon, but tens of satellites would be required for prompt attack of one target. For example, the constellation considered in Section 9, which could attack any point on the earth within about 30 minutes, would require nearly 100 satellites. If the promptness requirement was relaxed to a 45-minute response time, roughly 50 satellites would still be required.

For comparison, a missile capable of putting a given mass into low earth orbit can deliver the same mass to a range of 20,000 kilometers—halfway around the Earth. The flight time would be roughly 45 minutes. This one ballistic missile could therefore provide global coverage with the same response time as a constellation of 50 satellites requiring 50 comparable missiles to launch them into space.

For the five nuclear weapon states, the relative cost of a space-based system would be even higher, because they already possess intercontinental range ballistic missiles that could provide a prompt ability to attack ground targets globally.

Reliability

Unlike ground-based weapons, space-based weapons that have been launched must remain operational without either routine or emergency maintenance. As a consequence, space-based weapons would be less reliable and an attacker would have less confidence in using them for an attack than ground-based missiles. If a space-based weapon in the proper position for an attack failed, other satellites in the constellation could be used in its place but could not meet the same promptness criterion, because they would take time to move into position.

3. Space basing is unsuited to ballistic missile defense using kinetic interceptors. Not only would such a system be expensive, its intrinsic vulnerabilities would allow an attacker to readily negate its defensive capability. (Section 9)

The global coverage space-based weapons can provide is also a key motivation for deploying ballistic missile defense interceptors in space. The United States is conducting research on various types of missile defense systems designed to attack long-range ballistic missiles during their boost phase (the time when the rocket booster is still burning), which lasts only minutes. The short time available means that interceptors must be located close to the launch site of the missile; against large countries it may not be possible to deploy ground or air-based interceptors close enough. In contrast, a constellation of spacebased interceptors in low earth orbits could provide global coverage. Thus, in principle, a space-based boost-phase missile defense system could offer capabilities that would not be available with a ground- or air-based system.

However, because of the short response time this mission requires, the system would be intrinsically vulnerable to debilitating attack and to being overwhelmed. Any country with the capability to launch a long-range ballistic missile could also develop an effective capability to destroy satellites in low earth orbit using ASATs launched on short-range missiles. Once one or more space-based interceptors were destroyed, producing a hole in the defense constellation, an attacker could launch a long-range missile through this hole. If the defense used one of its interceptors to protect itself, it would still remove the interceptor from the constellation and create a hole.[2]

2. Instead of designing an ASAT weapon, an attacker could launch a ballistic missile without a warhead, thus forcing the defense to waste one of its interceptors and create a hole.

Alternatively, an attacker could overwhelm the defense. A defense system designed to intercept one ballistic missile launched from any given region would require many hundreds or even a few thousand orbiting interceptors, depending on the design of the constellation and the interceptors. Increasing the defense capability so the system could attack two missiles launched simultaneously from the same region would require doubling the total number of interceptors in the constellation. Because the system costs would increase rapidly with the number of interceptors, any plausible defense system would be designed to intercept only one or two ballistic missiles launched simultaneously. Thus, any country launching more than one or two missiles roughly simultaneously from the same region would penetrate such a defense, even if it worked perfectly.

4. A nation could not use space-based weapons to deny other countries access to space, although it could increase the expense of such access. (Section 9)

At first glance, it might appear that the first country to deploy a system of space-based missile defense interceptors with global coverage could control access to space by intercepting space launch vehicles, which are similar to long-range missiles. However, as discussed in Point 3, the vulnerabilities of a space-based missile defense system would render it ineffective at prohibiting satellite deployments by other countries.

Any country that can launch satellites has the technical capability to attack space-based interceptors to create a hole in the constellation. While requiring a country to attack the defense system prior to launching a satellite would increase the cost of placing a satellite in space, it could not in this way deny the country the ability to launch satellites.

5. Ground-based and space-based ASATs have relative advantages and disadvantages depending on the method of attack. Jamming or dazzling attacks would only be practical using ground-based ASATs, whereas high-power microwave attacks would only be practical from space. Kinetic energy attacks could be conducted from the ground or space; in this case, space-based ASATs may offer advantages for very prompt

attacks or for simultaneous attacks on numerous satellites, but would be less reliable than ground-based ASATs because they must operate without maintenance. (Sections 11, 12)

Ground- or space-based ASATs can be used to damage or interfere with satellites. However, some means of attack are only feasible or practical using one or the other. For those types of attacks that could be conducted from both ground and space, an attacker's preference would depend on a wide range of factors, including the scale and time requirements of the attack, whether the attack is to be covert, and to what extent cost is a factor (keeping in mind that only space-faring nations would be capable of deploying space-based ASATs).

A trailing space-based kinetic energy ASAT (which would be deployed in the same orbit as the target satellite to trail it) could carry out an attack almost instantaneously from the time a decision was made to attack, assuming the attacker was able to communicate with the ASAT. This would not necessarily be the case for either space-based ASATs that are designed to attack satellites in other orbits or for ground-based ASATs: both types of ASATs may need to wait hours for the target satellite to be within range of attack. Similarly, it would take several hours for a ground-based kinetic ASAT to reach a target in geosynchronous orbit. Moreover, by deploying multiple trailing ASATs, an attacker could destroy or interfere with multiple satellites essentially simultaneously. On the other hand, space-based ASATs have some significant disadvantages compared with ground-based ASATs. Placing ASATs in space is much more costly than basing them on the ground. Once in space, these systems must work without maintenance, and reliability becomes an issue.

6. Satellites are intrinsically vulnerable to attack and interference. However, satellite systems can be designed to be less vulnerable than the individual satellites that compose the system. Moreover, air- and ground-based backup systems can provide some of the militarily relevant, time-urgent capabilities that would be lost if the satellite system was disrupted or destroyed. (Sections 10, 11, 12)

Because there is no place to hide in space, satellites are inherently vulnerable to interference and direct attack. However, steps can be taken to reduce the vulnerability of the system, including hardening satellite components, employing antijamming techniques, building redundant ground stations, developing the capability to quickly replace satellites, and distributing the task of a single satellite among clusters of smaller satellites. The commercial communications satellite industry routinely deals with the failure of satellites. It places spare satellites in orbit to allow rapid substitution when satellites fail, and can reroute communication traffic around a failed satellite.

Moreover, for many military missions, ground- and air-based components can serve as a backup on a regional rather than global level. The United States is not only the country most reliant on space-based systems, but also the one most capable of building alternative air- and ground-based backup systems.

7. Deploying defensive "bodyguard" satellites to protect other satellites against ASAT attacks cannot provide confidence in the survivability of those satellites. Doing so will therefore not preclude the need to take into account the vulnerability of satellite systems and to have backup systems for any essential military capability the satellites provide. (Section 11)

As discussed above, satellites are vulnerable to many types of attack and defending them is inherently difficult. Equally important, nations will not be able to rely on bodyguard satellites to protect their satellites from direct attack or interference by a determined adversary, because the limited amount of realworld testing that would be feasible would provide little confidence in the capability of the bodyguards.

8. No country can expect to have a monopoly on deployed ASATs. Space-faring nations have an inherent capability to deploy effective ASATs. Many other countries have the capability to develop means to destroy satellites or disrupt their performance, although the options of these countries will be limited relative to those of space-faring nations. (Sections 11, 12)

The technology required to build effective ground- and space-based ASATs is within the capability of any space-faring nation. These countries have the ability to place objects in orbit or lift them to geosynchronous altitude, to track objects in space, and to develop homing interceptors. They could develop systems to attack satellites in geosynchronous orbits as well as low earth orbits. They could also deploy ASATs relatively quickly in response to the deployment of ASATs by another country.

Other countries also have anti-satellite capabilities. Satellites in low earth orbit can be reached by ground-based ASATs using missiles that are much less capable than the launchers needed to deploy the satellites. Countries with short- or medium-range missiles can reach satellites in low earth orbit at an altitude of roughly half the range of these missiles. However, such countries would not necessarily have the ability to develop homing interceptors. For a destructive ASAT, these countries might therefore be limited to releasing clouds of pellets in the path of a satellite—a method whose effectiveness is uncertain and potentially limited.

It is also within the capability of many countries to use other methods of interference. For example, ground-based transmitters can be used to interfere with, or jam, satellite communications. While such jamming is unlikely to prevent a well-protected satellite from communicating, jammers can cause the satellite to use antijamming techniques that can significantly reduce the amount of information it can transmit. On the other hand, during a conflict such active methods of interference can be located and attacked.

Countries that possess both nuclear weapons and short- or medium-range missiles could explode a nuclear weapon in low earth orbit. The x-rays released would destroy unshielded satellites in low earth orbit that were in the line of sight of the explosion, and the explosion would generate persistent radiation that would last months to years and would damage unshielded satellites in low earth orbit. While high-altitude satellites would not be directly affected by the explosion, this radiation environment could make it more difficult for them to communicate with ground stations.

9. Several types of systems not designed as ASATs have an inherent ASAT capability. (Sections 9, 11)

Missile defense systems designed to intercept long-range ballistic missiles outside the atmosphere during the midcourse of their trajectory have significant ASAT capabilities against satellites in low earth orbits. These satellites orbit at altitudes similar to the altitude at which the defense is designed to intercept missiles. Unlike the case of an attacking missile, the trajectory and appearance of the satellite would be known in advance and the future trajectory would be predictable. Moreover, even highly controlled intercept tests of the defense against ballistic missile targets would provide confidence that the system would work against satellites, since the information provided to the defense about the missile target in these tests is comparable to what would be available in advance about satellite targets. Using interceptors against a satellite rather than against a missile warhead is also easier in that the attacker could take multiple shots at the satellite if the initial attack was not successful.

Defender satellites would also have an inherent ability to serve as ASATs. Such satellites would need to carry enough fuel to maneuver to intercept attacking ASATs, and this fuel could also provide the maneuvering capability to serve as a kinetic energy ASAT.

10. Being the first to deploy space-based weapons would not confer a significant or lasting military advantage. (Sections 5, 9, 11)

In ground wars, there may be clear advantages to being the first to occupy and exploit a strategic location. Digging in and preparing defenses may make it difficult for an adversary to take control of the area. There are no such strategic locations in space, and a defender will not have the kinds of advantages in space as on the ground. No foreseeable space-based technologies would allow one country to prevent another from deploying space weapons or would allow it to reliably protect its satellites.

Section - 3

Technical Implications and General Conclusions

This report discusses the implications of some of the basic technical issues that govern the use of space. Here we summarize for each section of the report some of the general conclusions that result from analysis of these issues.

Basics of Satellite Orbits (Section 4)

- The speed of a satellite is not arbitrary: it is determined by the satellite's orbit and is closely tied to the satellite's altitude.
- A satellite's orbit does not depend on its mass. All objects with the same velocity (speed and direction) at a given point in space follow the same orbit.
- Satellites close to the Earth move faster than those at higher altitudes and, when viewed from the ground, cross the sky faster. Satellites in low earth orbits (hundreds of kilometers above the Earth) move rapidly relative to the Earth, completing an orbit in 1.5 to 2 hours.
- Satellites in higher orbits move at slower speeds than those in lower orbits, and the distance they travel in one orbit is longer. As a result, the time required for a satellite to orbit (the orbital period) increases with altitude. Only one altitude (36,000 km) permits satellites to orbit at the same rate at which the Earth rotates; such satellites are called geosynchronous.
- Once in orbit, a satellite does not need constant powering to remain in flight, as airplanes do. Satellites use small onboard rocket engines to maneuver in space.

- A satellite's orbit always lies in a plane that passes through the center of the Earth. The angle between that plane and the plane of the equator is called the orbit's inclination.

Types of Orbits, or Why Satellites Are Where They Are (Section 5)

- Because the Earth rotates underneath the satellite as it orbits, a satellite in a polar orbit (an orbit that passes over both poles) travels directly over every point on Earth. Satellites in equatorial orbits only travel directly over the equator. Satellites may be in orbits with inclinations between these two extremes; in such cases, the satellite travels directly over points on the Earth with a latitude equal to or less than the satellite's inclination angle.
- Satellites that are in equatorial orbits and that have an orbital period of 24 hours stay fixed over a point on the equator; they are called geostationary. While geostationary orbit is useful for hosting communications and broadcasting satellites, it is not well suited to such missions as high-resolution imagery or ground attacks, because such an orbit requires a very high altitude (36,000 km). Furthermore, because geostationary satellites travel only in the equatorial plane, they have difficulty communicating with the Earth's polar regions.
- Satellites that are not in geostationary orbit move with respect to the ground, and so constant coverage of a particular location on the Earth requires a constellation of satellites.
- Satellites at high altitudes can see more of the Earth's surface at one time than can satellites at lower altitudes.
- Satellites that need to be close to the Earth to perform specific missions, for example, to take high-resolution images of the ground, must be located in low earth orbits. Being closer to the Earth's surface makes these satellites more vulnerable to interference from ground-based methods of attack.

Maneuvering in Space (Section 6) **and Implications of Maneuvering for Satellite Mass** (Section 7)

- Maneuvering a satellite, which requires changing the speed of the satellite or its direction of travel, can require a large expense of energy. The mass of propellant a satellite needs to change its velocity increases exponentially with the amount of velocity change. The difficulty and cost of placing large amounts of propellant in space therefore limit how much maneuvering satellites can do.
- Maneuvers to change the satellite's orbital plane can require large changes in the satellite's velocity and can therefore require large amounts of propellant. By contrast, maneuvers that alter the shape or altitude of the orbit but that do not change the orbital plane generally require much less propellant, especially if the satellite moves between low earth orbits.
- Propulsion using new technologies can generate substantially more velocity change per unit mass of fuel than conventional chemical propellants do. This reduces the mass of fuel a satellite needs to carry to perform a given maneuver. While more efficient, the new propulsion technologies that will be available in the foreseeable future cannot be used to carry out maneuvers quickly, which limits the tactical utility of these technologies.

Getting Things into Space: Rockets and Launch Requirements (Section 8)

- Placing an object in orbit is much more demanding than simply lifting it to a high altitude. Although short- and medium-range ballistic missiles can reach the altitudes of satellites, once there they cannot produce the high speeds necessary to put a satellite into orbit. Even a long-range (10,000 km) ballistic missile cannot put its full payload into orbit.
- A rule of thumb is that a ballistic missile that can deliver a given payload to a maximum range R on the Earth can lift that same payload vertically to an altitude R/2 above the Earth. Reducing the mass of the payload increases both R and R/2.

- Modern rockets can deliver into low earth orbit a payload that is only 2–4% of the total mass at launch. Roughly 45 tons of propellant are required for every ton of payload placed in orbit.
- How much mass a launch vehicle can place in orbit depends on the location of the launch site and the intended orbit. Since the rotational speed of the Earth's surface is largest near the equator, launching from sites near the equator allows the launcher to take advantage of that additional speed.
- Reducing the size and mass of satellites can reduce launch costs and may allow satellites to hitch a ride on other launches, which can be cheaper than using a dedicated launcher and may be scheduled more quickly.

Space Basing (Section 9)

- Operating in space has a number of important consequences: First, placing satellites in orbit is costly. Second, satellites in low earth orbits move relative to the Earth's surface, leading to an intrinsic problem with absenteeism (i.e., low-earth-orbiting satellites spend most of their time above the wrong part of the Earth), and missions that require low-altitude satellites to be at a specified location require multiple satellites in orbit. Third, repairing, refueling, or updating satellites in orbit is difficult and costly, so it is rarely done. As a result, the reliability of a space-based system decreases with time. Fourth, the space environment is relatively hostile to satellites, with high levels of radiation, large temperature changes from sun to shadow, and fast-moving space debris.
- Space-based ground-attack weapons intended for prompt, ondemand attacks and global reach would have a high absentee ratio, since a large number of satellites would be needed to ensure that one is in place for the mission at all times. Ground-based systems could provide these capabilities on the same timescale, with greater reliability and at a cost many tens of times less.
- Basing a boost-phase missile defense system in space is suggested as a means to cover missile launches anywhere

in the world. Because the response time required for boost-phase missile defense is very short, a constellation of hundreds or thousands of space-based interceptors would be required. In addition, the interceptors would need to be able to maneuver significantly when attacking a missile, and the propellant required for these maneuvers would drive up their mass.

- Because of the short response time required, space-based interceptors would be placed in low orbits, where they would be vulnerable to attacks by short-range missiles. Because interceptors must be close to the launch site of a missile to have time to reach the missile, destroying several interceptors could create a hole in the constellation through which an attacker could fire a longrange missile.
- Some missions discussed for the military space plane, such as releasing multiple satellites in different orbits, would require significant maneuverability in space and therefore large masses of propellant. For missions other than those within the same orbital plane, it may be more efficient to launch a new vehicle for each satellite rather than to maneuver the vehicle in space to release multiple satellites.
- For missions that require maneuvering and that use satellites with low mass, such as simple inspector satellites, the mass of propellant needed for the mission may not be prohibitively large.

Elements of a Satellite System (Section 10)

- A satellite is made up of a number of different elements, including solar panels, payloads, and communications devices. The system also includes ground stations to control the satellite and communications equipment for linking with the satellite. All these elements can be targeted to interfere with a satellite system.
- Satellites vary greatly in size, with masses from a few kilograms to a few tons.

Overview of Interfering with Satellite Systems (Section 11)

- Interference can range from temporary or reversible effects to permanent disabling or destruction of the satellite. Many methods can be used to interfere with satellites, including electronic interference with communication systems, laser interference with imaging sensors, laser heating of the satellite body, high-power microwave interference with electrical components, collision with another object (kinetic-kill), and nuclear explosions.

- Because satellites can be tracked and their trajectories can be predicted, they are inherently vulnerable to attack. However, a satellite's vulnerability to ASAT attack does not guarantee the effects of an attack will be predictable or verifiable, and this may limit the ASAT attack's usefulness.
- Jamming satellite ground stations (the downlinks) and the satellite's receivers (the uplinks) is relatively simple to do on unprotected systems such as commercial communications satellites. Jamming protected systems, such as military communications satellites, is much harder. An adversary need not be technologically advanced to attempt a jamming attack.
- Ground-based lasers can dazzle the sensors of high-resolution reconnaissance satellites and inhibit observation of regions on the Earth that are kilometers in size. With high enough power, ground- and space-based lasers can partially blind a satellite, damaging relatively small sections of the satellite's sensor.
- A high-power laser can physically damage a satellite if its beam can be held on the satellite for long enough to deposit sufficient energy. This can result in overheating the satellite or damaging its structure.
- High-power microwave weapons can disrupt or damage the electrical systems of a satellite if enough of their energy enters these systems. Such attacks would be conducted from space rather than from the ground. Microwave attacks could attempt to enter the satellite through its antennae (a front-door attack) or through other routes, such as seams in the satellite's casing (a back-door attack). The effectiveness of both types of attack would be difficult to predict.

- Satellites in low earth orbits can be attacked by kinetic-kill ASATs carried on short-range missiles launched from the ground. ASATs stationed on the ground or in low earth orbits can be designed to reach targets at higher altitudes in a matter of hours.
- A nuclear explosion at an altitude of several hundred kilometers would create an intense electromagnetic pulse that would likely destroy all unshielded satellites that are in low earth orbit and in the line of sight of the explosion. In addition, persistent radiation created by the explosion would slowly damage unshielded satellites at altitudes near that of the detonation.

Topics in Interfering with Satellites (Section 12)

- Space-based ASATs are likely to be deployed in one of four ways: co-orbital with and a short distance behind the target satellite (a trailing ASAT); attached to the target (sometimes called a para-sitic ASAT); in a distant part of the same orbit, requiring a maneuver to approach and attack the target; or in a crossing orbit, keeping its distance from the target until the time of engagement. Different interference methods would be suited to different deployment configurations.
- To be covert, a space-based ASAT must elude detection and/or identification during launch, during deployment maneuvers, and while in orbit. No country could assume its deployment of a space-based ASAT would remain covert. At the same time, no country can assume it would be able to detect or identify a spacebased ASAT deployed by another country. Detecting a covert weapon may allow the targeted country to publicly protest its presence and to prepare tactical alternatives to the targeted satellite, but may not guarantee the country's ability to defend against the ASAT.
- A simple anti-satellite weapon that could be used by an attacker with relatively low technical sophistication is a cloud of pellets lofted into the path of a satellite by a short- or medium-range ballistic missile. The effectiveness of

such an attack would depend on the attacker's ability to determine the path of the target satellite with precision and to control its missile accurately. Unless the attacker can do both, such an ASAT would have limited effectiveness.

- Many systems that rely on satellites can be made to withstand interference that disrupts an individual satellite. The consequences of an attack on a satellite in the system can be reduced by smart design, including building in redundancy, adding backup systems and spares, and developing alternative means to perform vital functions.

Section - 4

The Basics of Satellite Orbits

Motion in Space Vs Motion in the Atmosphere

The motion of objects in the atmosphere differs in three important ways from the motion of objects in space. First, the speed of an airplane bears no particular relationship to its flight path or altitude: two airplanes can follow the same flight path at different speeds. In contrast, a strict relationship holds between a satellite's orbit and its speed: as an example, for circular orbits, all satellites traveling at the same altitude must have the same speed, and satellites at different altitudes must have different speeds. As detailed below, this relationship between altitude and speed severely constrains the behavior of objects in space.[1]

Second, an airplane uses the presence of the air not only to stay aloft, but to maneuver. Similar to the way a boat maneuvers in water by using oars or a rudder to push against the water, an airplane uses wing flaps and a rudder to change direction by pushing against the air. In the vacuum of space, this is not possible. A satellite must instead use small rocket engines to maneuver. Because such rockets require propellant, this has important implications for the design and capabilities of satellites.

Third, since air resistance to an airplane's motion continually slows it down, an airplane must be continually powered by engines to stay in flight. This is not true for a satellite in the vacuum of space. A rocket booster is needed to place a satellite in orbit,

1. Airplanes and satellites resist gravity in very different ways. An airplane stays aloft because the motion of its wings through the air creates lift forces; we discuss the physics behind satellites below.

but once there it circles the Earth in its orbit without requiring constant powering from rocket engines.[2] The Moon, for example, is a naturally occurring satellite of the Earth that continues to orbit the Earth without benefit of a rocket booster.

Orbital Basics

This section discusses, in general terms, the physics of satellite orbits. It summarizes the key concepts of orbital mechanics, which define the properties of a satellite in orbit: its orbital speed, orbital period, and the orientation of the orbit with respect to the equator. Mathematical details are provided in the Appendix to Section 4.

In general, satellite orbits are ellipses. However, this section first considers the special case of circular orbits before turning to elliptical orbits. Circular orbits are simpler to describe and understand, and they are used for many applications.

Orbital Speed of Satellites

For a satellite in a circular orbit, the relationship between the orbital speed and altitude is strict. The task of the rocket launching the satellite is to release the satellite at the appropriate place in space, with the appropriate speed and direction of motion to put it in the desired orbit.

How a satellite stays in orbit may be thought about in two equivalent ways, both of which explain the relation between the satellite's altitude and speed.

The satellite's motion may be seen as creating a centrifugal force that opposes the attraction of gravity. For example, imagine attaching an object to a string and swinging it in circles. The object pulls outward against the string, and that outward force (the centrifugal force) becomes greater the faster the object swings. At the proper speed, the centrifugal force of the satellite due to its motion around the Earth just balances the pull of gravity, and the satellite remains in orbit.

Since the gravitational pull grows weaker the further a satellite is from Earth, the centrifugal force needed to balance

2. Satellites in orbits up to altitudes of several hundred kilometers will experience slight atmospheric drag, thus they must periodically fire rocket engines to counteract that drag and remain in orbit.

gravity also decreases with distance from the Earth. The higher the satellite's orbit, the lower its orbital speed. Alternately, the satellite can be viewed as constantly falling toward the center of the Earth. However, since the satellite is also moving parallel to the Earth's surface, the Earth continually curves away from satellite. In a circular orbit, the satellite's speed is exactly what is required so that it continually falls but keeps a constant distance from Earth. The required speed depends on the satellite's altitude because of the Earth-satellite geometry and because the rate at which the satellite falls toward the Earth depends on the strength of gravity at its altitude. For the case of a satellite traveling in a circular orbit, Figure 4.1 and Table 4.1 show the orbital speed for various altitudes. The required speeds are very high: satellites in low altitude orbits (up to about 1,000 km) travel at 7 to 8 km/s. This speed is roughly 30 times faster than a large passenger jet.[3] (By comparison, the rotational speed of the Earth's surface at the equator is 0.46 km/s.) If an object is launched from Earth with a speed of 11.2 km/s or greater, Earth's gravity is not strong enough to keep it in orbit and it will escape into deep space. This speed is therefore called the escape velocity.

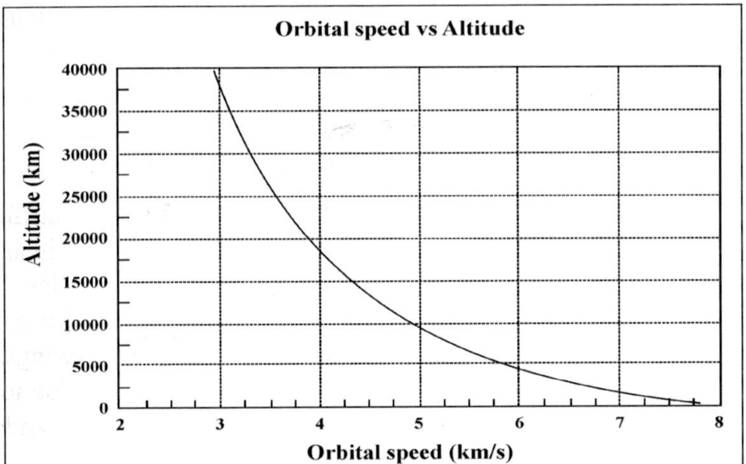

Figure 4.1. Orbital speed for satellites in circular orbits at different altitudes.

3. A Boeing 747 aircraft has a typical cruising speed of about 900 km/hr, or 0.25 km/s; see "747 Technical Specifications," http://www.boeing.com/commercial/747family/technical.html, accessed January 15, 2005.

Table 4.1. Selected values for the speed and altitude of satellites in circular orbits.

Altitude (km)	Orbital Speed (km/s)
200	7.8
500	7.6
1,000	7.4
5,000	5.9
10,000	4.9
Semisynchronous: 20,200	3.9
Geosynchronous: 35,800	3.1

Note that the speed needed to keep a satellite in orbit does not depend on the mass of the satellite. This is fundamental to understanding issues related to space: the trajectory of an object in the vacuum of space does not depend on its mass. This means that lightweight debris or even paint chips will move on the same trajectory as a heavy satellite, and that a heavy satellite and a lightweight satellite (a micro-sat) with the same velocity will travel on the same orbit.[4] Our intuition on this point tends to be clouded by the fact that on Earth, air resistance affects the motion of light objects more than heavy objects. As noted above, once a satellite has been accelerated up to orbital speed by a rocket, it does not need to be continually powered to stay in orbit. This is a consequence of Newton's first law of motion, which says that in the absence of forces such as friction and air resistance, an object at rest will stay at rest and an object in motion will stay in motion. As a result, once put in motion by a rocket, a satellite will stay in motion, with the Earth's gravity bending its path from a straight line into an orbit.

This means that satellites can stay in orbit for long periods of time, since they do not need to carry large amounts of fuel to keep them moving. It also means that once in orbit, other objects—such as empty rocket stages, screws from the mechanisms that release satellites from the rocket that put them in space, and other debris—will stay in orbit essentially indefinitely, unless they are

4. An important consequence of this fact for missile defenses that are designed to intercept above the atmosphere is that a heavy warhead and lightweight decoys will follow the same trajectory and, therefore, cannot be distinguished by observing their trajectories.

at low enough altitudes that atmospheric drag slows them over time and they fall to Earth. This is the essence of the space debris problem: once debris is in orbit it will remain there and thus the amount of such debris will accumulate over time. Without efforts to minimize the creation of debris, some regions of space could eventually contain so much orbiting debris that it would be difficult to operate satellites there without the risk of collisions.

Orbital Period of Satellites

Another key parameter used to describe satellites is the time it takes for the satellite to travel around the Earth once, that is, to complete one orbit. This time is known as the period of the orbit. Since as the altitude of the orbit increases the satellite both moves more slowly and must travel farther on each orbit, the period increases with the altitude of the orbit.

Table 4.2 and Figure 4.2 show the period for circular orbits at various altitudes. For low altitude orbits (a few hundred kilometers altitude), the period is about 90 minutes; at higher altitudes, the period increases.

Since one day is roughly 1,440 minutes, the plot shows that a satellite at an altitude of about 36,000 kilometers orbits once a day—at the same rate the Earth rotates. Such orbits are termed geosynchronous. A satellite placed in a geosynchronous orbit above the equator is unique in that it stays over the same point on the Earth. Such geostationary orbits have important uses, as discussed in the next section.

Table 4.2. Select values for the orbital periods and altitudes of satellites in orbit.

Altitude (km	Orbital Period (minutes)
200	88.3
500	94.4
1,000	104.9
5,000	201.1
10,000 3	47.4
Semisynchronous: 20,200	718.3 (12 hours)
Geosynchronous: 35,800	1436.2 (24 hours)

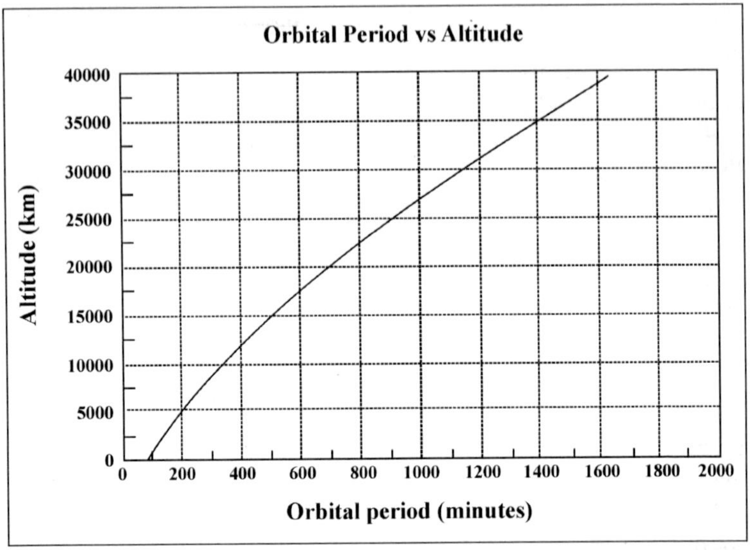

Figure 4.2. Orbital period as a function of altitude for circular orbits.

Orientation of the Plane of the Orbit A satellite's orbit always lies in a plane, and that plane passes through the center of the Earth. Describing a satellite's orbit requires specifying the orientation of this orbital plane. When the plane of the orbit includes the Earth's equator, the orbit is known as an equatorial orbit. In general, the plane of the orbit lies at an angle with respect

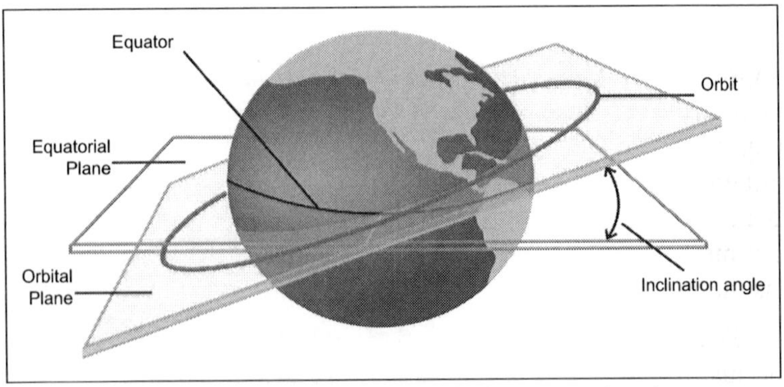

Figure 4.3. Inclination angle of orbits. Note that the ground track of a satellite, which is the line of points on the Earth directly below the satellite, cannot reach above a latitude equal to the inclination angle of the orbit.

to the Earth's equatorial plane; that angle is called the inclination angle (see Figure 4.3).

When the inclination angle is 90 degrees, the orbital plane contains the Earth's axis and the orbit passes over the Earth's poles. Such an orbit is known as a polar orbit.

The inclination of the orbit determines what parts of the Earth the satellite travels over. The path on the Earth's surface that lies directly beneath the satellite is called the ground track of the satellite. Figure 4.3 shows that a satellite in an orbit with inclination near zero passes over only those parts of the Earth that lie in a narrow band around the equator. Therefore, a satellite with inclination near zero may not be able to observe or communicate with parts of the Earth near the poles. More generally, Figure 4.3 shows that a satellite in an orbit with inclination angle θ cannot pass directly over any location on Earth with latitude greater than θ. (Recall that the latitude of a point on Earth is the angular position north or south of the equator.)

A satellite launched from a site at latitude θ follows an orbit with inclination greater than or equal to θ. From a launch site at latitude θ it is not possible to launch a satellite into an orbit with inclination less than θ. As a result, a launch site that is not on the equator cannot place a satellite directly into an equatorial orbit. To change its inclination after it is in orbit, a satellite would need to maneuver, which requires propellant.[5]

Elliptical Orbits In general, an orbit is not a circle, but an ellipse. A circle is the set of all points equidistant from a given point, which is the center of the circle. Instead of a center, an ellipse has two foci. The ellipse consists of those points with the property that the sum of the distance from each point to the two foci is constant. (So a circle is the special case in which the two foci merge to become a single point.) An elliptical satellite orbit has the Earth at one of the foci (see Figure 4.4).

The line extending across the widest part of the ellipse is called the major axis; it passes through both foci. The line extending

5. An orbit with inclination θ will be achieved if the satellite is placed in orbit with a velocity pointing due east or west when it is above the launch site (at latitude θ). If the satellite has an orbital velocity in any other direction at that location, it will pass over latitudes greater than that of the launch site, and the inclination angle will be greater than θ.

across the narrow part of the ellipse, perpendicular to the major axis, is called the minor axis (see Figure 4.4).

The amount by which an ellipse deviates from a circle is described by its eccentricity, which varies between zero (corresponding to a circle) and one (corresponding to an infinitely thin ellipse).

On an elliptical orbit, the altitude and the speed of the satellite vary with position around the orbit. The point on an elliptical orbit

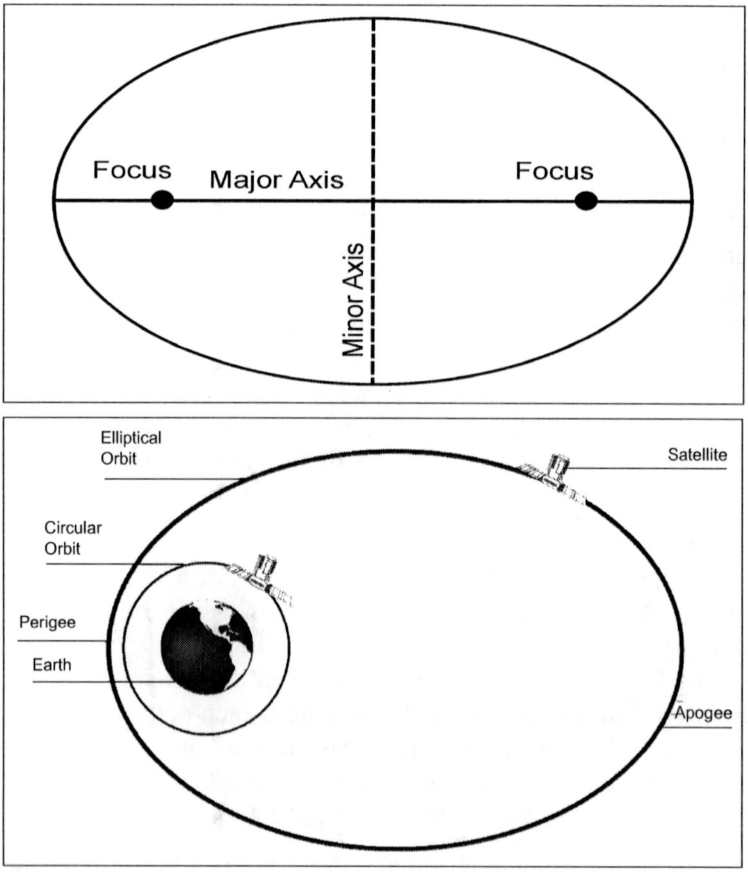

Figure 4.4. The top figure shows an ellipse, with the major and minor axes and the two foci marked. The lower figure shows satellites on circular and elliptical orbits, with the Earth at the center of the circular orbit and at one of the foci for the elliptical orbit.

at which the satellite is closest to the Earth is called the perigee of the orbit, and the point at which it is furthest from the Earth is called the apogee. The perigee and apogee lie at opposite ends of the major axis.

A satellite on an elliptical orbit moves faster when it is closer to the Earth (near perigee) and more slowly when it is farther from the Earth (near apogee). The speed of a satellite at a given point depends not only on its altitude at that time, but also on the shape of the orbit (in particular, on the length of the major axis). If a satellite is on an elliptical orbit, its speed at a given altitude can be either higher or lower than the speed the satellite would have on a circular orbit at the same altitude, depending on the shape of the ellipse (see the Appendix to Section 4 and Section 6).

The period of the orbit also depends on the length of the major axis: it increases as the major axis increases. Elliptical orbits can be geosynchronous but not geostationary since the satellite's orbital speed varies with time. (For more details on elliptical orbits, see the Appendix to Section 4.)

Appendix - 4

Circular and Elliptical Orbits

Circular Orbits

For the simple case of a satellite traveling in a circular orbit at altitude h with speed V, the centrifugal force equals the gravitational force on the satellite:

$$\frac{mV^2}{(R_e+h)} = \frac{GmM_e}{(R_e+h)^2} \tag{4.1}$$

where m is the mass of the satellite, G is the gravitational constant, Me is the mass of the Earth (GMe = 3.99 × 1014 m3/s2), and Re is the average radius of the Earth (6,370 kilometers). Thus the speed of the satellite is related to its altitude by the simple equation[1]

$$V = \sqrt{\frac{GM_e}{h+R_e}} \tag{4.2}$$

It is useful to let r be the distance from the satellite to the center of the Earth, so that

$$r \equiv R_e + h \tag{4.3}$$

Using this notation, Equation 4.2 can be written

$$V = \sqrt{\frac{GM_e}{r}} \tag{4.4}$$

Notice that the mass of the satellite does not appear in Equations 4.2 or 4.4. The period can be found by dividing the distance the satellite travels in one orbit (in this case, the circumference of a

1. This discussion assumes the Earth is spherical, with a radius Re and mass Me. The implications of the lack of spherical symmetry are discussed in Section 5.

circle with radius h + Re) by the speed of satellite, which is given in Equation 4.2. The period Pcirc of a circular orbit is given by

$$P_{circ} = 2\pi\sqrt{\frac{(h+R_e)^3}{GM_e}} = 2\pi\sqrt{\frac{r^3}{GM_e}} \tag{4.5}$$

Elliptical Orbits

An ellipse surrounds two points called foci. An elliptical satellite orbit has the Earth at one of the foci.

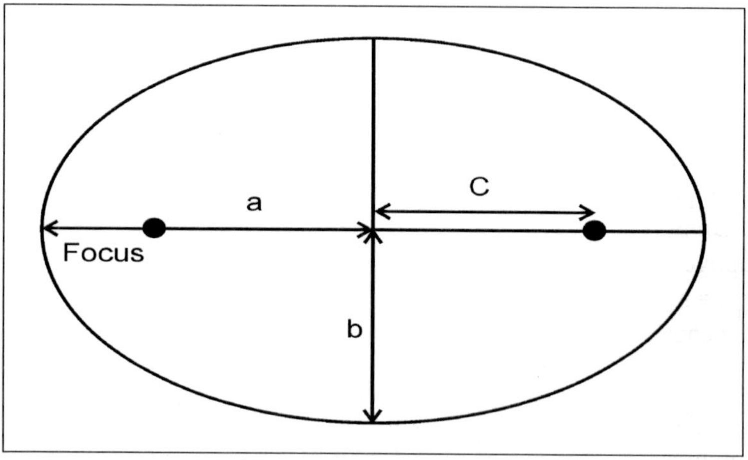

Figure 4.5. This figure shows an ellipse with major axis of length 2a, minor axis of length 2b, and distance between the foci equal to 2c.

The line containing the two foci is the major axis: its length is typically labeled 2a. The minor axis, the line perpendicular to it through the center of the ellipse has a length labeled 2b. The distance between the foci is called 2c (see Figure 4.5). These quantities are related by

$$a^2 = b^2 + c^2 \tag{4.6}$$

The perigee of the orbit is the point at which the satellite is closest to the Earth; the distance from perigee to the center of the Earth is denoted by r_p. Similarly, the apogee is the point at which the satellite is farthest from the Earth, and the distance from the center of the Earth to the apogee is denoted by r_a. From the geometry, it is clear that

$$r_a + r_p = 2_a \text{ and } r_a - r_p = 2c \tag{4.7}$$

Since $a = (r_a + r_p)/2$, a can be interpreted as the mean distance of the orbit from the center of the Earth. The amount by which an ellipse deviates from a circle is quantified by the eccentricity, e, which has values between zero (corresponding to a circle) and one (corresponding to an infinitely thin ellipse). The eccentricity is proportional to the distance between the foci, and is given by

$$e = \frac{c}{a}\sqrt{1 - \frac{b^2}{a^2}} \tag{4.8}$$

$$e = \frac{r_a - r_p}{r_a + r_p} \tag{4.9}$$

Most satellites are in orbits that are roughly circular, so that it is possible to talk about "the altitude" of the satellite. As an example, consider an orbit with an altitude at perigee (h_p) and apogee (ha) of 500 and 800 kilometers, respectively. In this case, r_p ($= h_p + R_e$) and r_a ($= h_a + R_e$) differ only by about 4%, and the eccentricity is 0.02. It can also be shown that

$$r_a = a(1 + e) \text{ and } r_p = a(1 - e) \tag{4.10}$$

Finally, conservation of angular momentum of the satellite requires that $r_a V_a = r_p V_p$, where V_a and V_p are the speeds of the satellite at apogee and perigee, respectively. From this it follows that

$$\frac{r_a}{r_p} = \frac{V_p}{V_a} = \frac{1+e}{1-e} \tag{4.11}$$

The speed of a satellite at a point on an elliptical orbit depends on its altitude h at that point, and is given by[2]

$$V = \sqrt{GM_e\left(\frac{2}{h+r_e} - \frac{1}{a}\right)} = \sqrt{GM_e\left(\frac{2}{r} - \frac{1}{a}\right)} \tag{4.12}$$

This reduces to Equation 4.2 for circular orbits, since in that case $a = h + R_e$, the radius of the orbit. This equation shows that a satellite on an elliptical orbit moves faster when it is closer to the Earth (near perigee) and more slowly when it is farther from the Earth (near apogee). Kepler showed, in the early 1600s, that a satellite moves in such a way that a line drawn from the center of

2. David Vallado, Fundamentals of Astrodynamics and Applications, 2nd ed. (El Segundo, CA: Microcosm Press, 2001), 27.

the Earth to the satellite sweeps out equal areas in equal time. This general property is a straightforward consequence of conservation of angular momentum. The period of an elliptical orbit is[3]

$$P = 2\pi\sqrt{\frac{a^3}{GM_e}} \tag{4.13}$$

which is Equation 4.5 with the radius r = (h + R^e) of the circular orbit replaced by the semi major axis a of the ellipse.

3. Vallado, 30.

Section - 5

Types of Orbits, or Why Satellites Are Where They Are

The choice of a particular orbit for a satellite depends mainly on its mission. For example, a remote-sensing satellite that collects high-resolution images of the Earth's surface should be as close to the Earth as practical. Consequently, such satellites are in low earth orbits. A commercial broadcast or communications satellite has other requirements. It should be able to send and receive signals from a large geographic area. It should preferably be in a fixed location, so ground stations will not need expensive satellite-tracking equipment. For these reasons, most communications satellites are in equatorial geostationary orbits. The orbits for other satellites will similarly be chosen based on their missions.

This tight correlation between mission and orbit has important consequences. For example, although the traditional notion of "territory" as a fixed ground area or fixed volume of airspace is not relevant in space (since all permanent residents must be orbiting), a few special orbits are uniquely suited to a specific purpose and are therefore highly valuable. As a result, it is often possible to guess at the function of an unknown satellite by observing what orbit it follows.

In this section, we first discuss several important characteristics of orbits. These include those proscribed by geometry: the motion of satellites with respect to the Earth, the elevation angle above the horizon of the satellite for different positions on the ground, the maximum ground area that a satellite can observe, and the time it takes for a transmission signal to travel between the ground and satellite. Other orbital characteristics are a consequence of the local environment, such as radiation and atmospheric effects. We then discuss the most common types of orbits and the satellites that populate them.

The Constraints of Geometry

The Motion of Satellites Relative to the Surface of the Earth

In addition to the orbital motion of the satellite, the Earth also spins on its axis, and the motion of the satellite relative to the surface of the Earth is determined by both effects. Because of the rotational motion of the Earth, when a satellite returns to the same point in its orbit one period later, it is no longer over the same location on the Earth—unless the orbit is chosen so that the satellite's period is one day. In this case, the satellite's orbital period is the same as the Earth's rotational period, and such orbits are called geosynchronous. Geosynchronous orbits can be circular or elliptical and can have any inclination angle. As shown in Section 4 (see Figure 4.2 and Table 4.2), circular orbits with an altitude of about 36,000 km are geosynchronous.

A circular geosynchronous orbit that lies in the equatorial plane (inclination of 0°) is a special case: it is geostationary. A satellite in this orbit stays fixed relative to the surface of the Earth and remains directly over a point on the equator. To an observer on the ground, the satellite appears motionless. Note that only in an equatorial orbit is it possible for a satellite to remain stationary over a point. Because there is only one geostationary orbit, space in this orbit is valuable.

Now consider the case of a satellite in orbit at several hundred kilometers altitude. As shown in Figure 4.2 and Table 4.2, such satellites have orbital periods of roughly 90 minutes. In 90 minutes, the surface of the Earth at the equator rotates about 2,500 km (at other latitudes, the distance rotated would be less than 2,500 kilometers). Thus, after one period, the satellite passes over the equator at a spot 2,500 km west of the spot it passed over on its previous orbit.

A satellite in a low-altitude orbit also is in view of a given location for only a short time. To a person on the ground directly under the orbit, the satellite appears above the horizon on one side of the sky, crosses the sky, and disappears beyond the opposite horizon in about 10 minutes. It reappears after about 80 minutes,[1] but does not pass directly overhead (unless the observer is at one of the poles), since the Earth has rotated during that time.

As the satellite moves in its orbit, the point on the Earth directly beneath the satellite traces out a path called the satellite's ground track. (The ground track of a geostationary satellite is simply a point on the equator.)[2] Figure 5.1 shows an orbit with inclination of 45°; the shaded disc is that part of the orbital plane lying inside the orbit. The line where the orbital plane touches the Earth's surface would be the ground track of the satellite if the Earth did not rotate. This figure shows why the ground track, when drawn on a flat map of the Earth, appears as a curve that passes above and below the equator, as shown in the upper panel of Figure 5.2. Half of the orbit lies above the equatorial plane and half lies below. Note that the maximum latitude the ground track reaches (north and south) is equal to the inclination of the orbit.

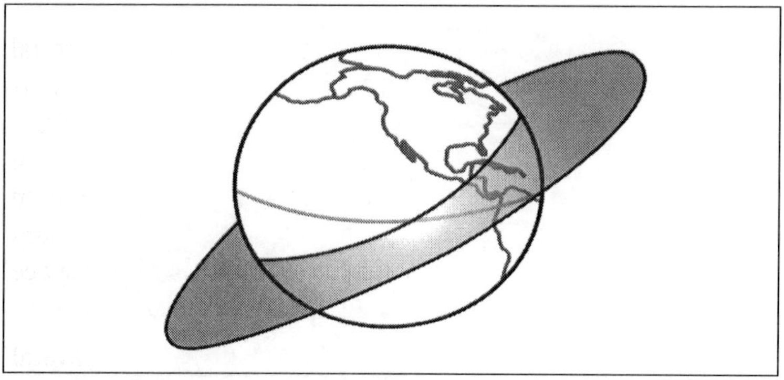

Figure 5.1. The shaded disc is that portion of the orbital plane lying within an inclined circular orbit. If the Earth did not rotate, the line where the shaded disc meets the Earth's surface would be the ground track of the satellite. This figure shows that, for an inclined orbit, the ground track of the satellite passes above and below the equator.

---

1. If it does not pass overhead on the subsequent orbit, it will not be visible unless it is at high enough altitude.

2. For a satellite in a geosynchronous orbit that is not geostationary, with a nonzero inclination θ, the ground track will be a figure eight centered at a fixed point on the equator. The top and bottom of the "8" will lie at latitude θ above and below the equator.

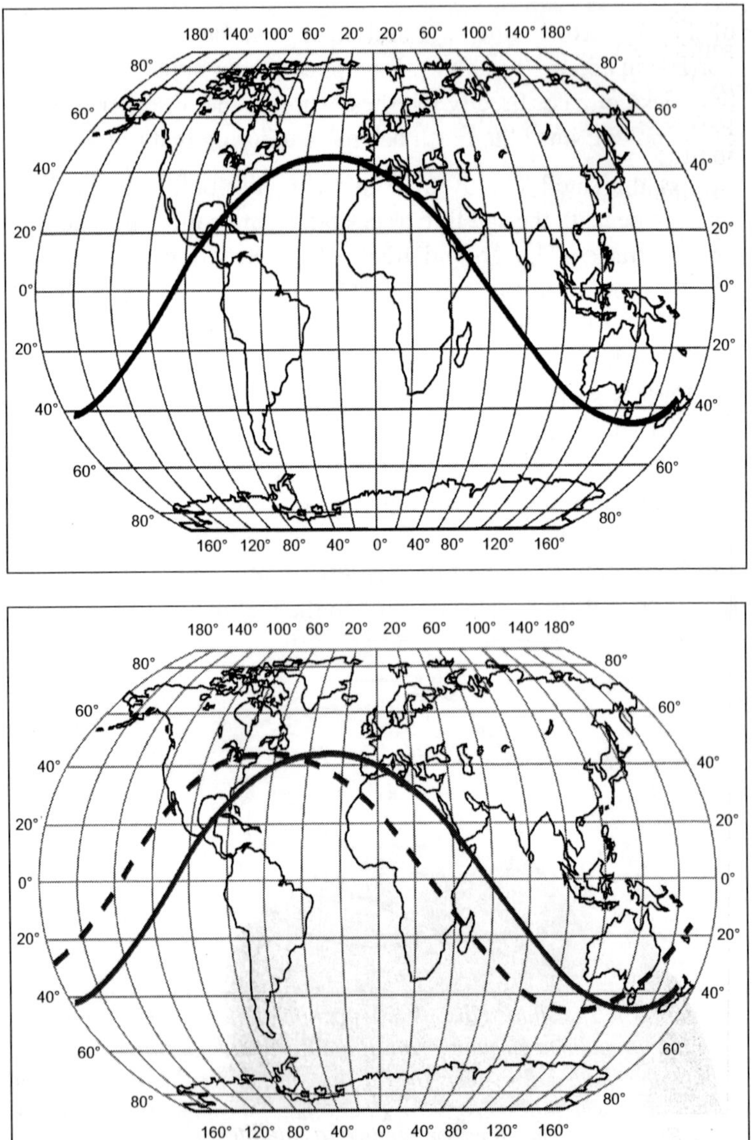

Figure 5.2. The upper panel shows the ground track for one pass of a satellite in a low earth orbit with an inclination angle of 45°. The track of the satellite on its next pass over this region is shown as the dashed curve in the lower panel. Since the Earth has rotated in the time the satellite was making one orbit, the ground tracks do not overlap.

Because the Earth rotates, the ground track does not lie in the same place on the Earth's surface during its next orbit, as shown by the dashed line in the bottom panel of Figure 5.2. Unless the period of the satellite is chosen to have a special value, the satellite in time flies over all points of the Earth between the maximum and minimum latitudes.

Elevation Angle of the Satellite: The elevation angle of a satellite is the angle between the satellite and the local horizon as seen from a particular location on the ground (Figure 5.3). It is a measure of how directly overhead a satellite is at a given time, with an elevation of 90° signifying the satellite is directly overhead. Because it is measured with respect to a specific ground location, it is different for different observers on the ground. It varies with time as the satellite moves through its orbit.

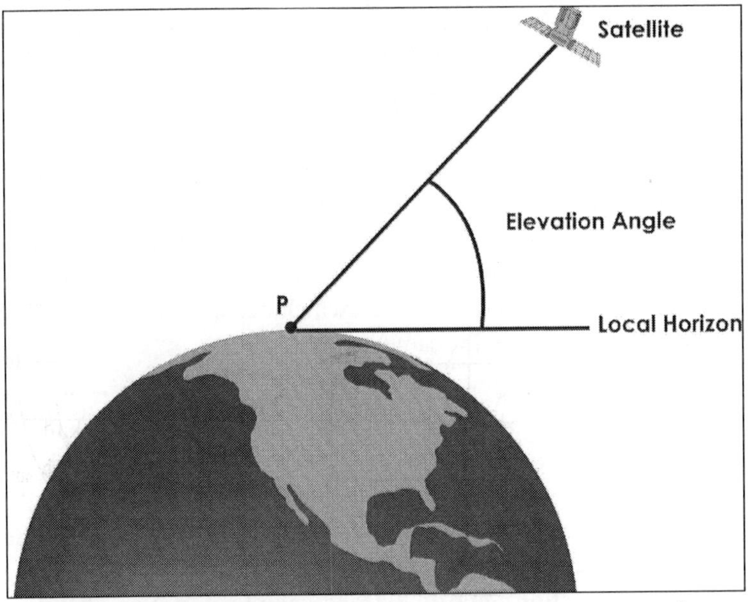

Figure 5.3. The elevation angle of a satellite at a given time for an observer at the point P on the Earth is the angle between the local horizon at P and a line from the observer to the satellite.

Specifically, the elevation angle at a given time depends on several parameters that describe the relative location of the satellite and the observer on the ground. These include the latitude and longitude of the observer, the altitude of the satellite, the inclination angle θ of the orbit, and where on the orbit the satellite is (the satellite's latitude and longitude). The exact relation is given in the Appendix to Section 5.

A few examples can provide a sense of how the observer's latitude, satellite altitude, and orbital inclination affect the elevation angle. Consider an observer at the equator (i.e., at a latitude of 0°) and a satellite in a circular equatorial orbit (i.e., at an inclination angle of 0°). As the satellite traverses its orbit, it passes directly over the observer, and its elevation angle increases from 0° to 90° and then decrease back to 0° for an observer at any point on the equator.

The satellite has an elevation angle of 90° only for points on the ground directly beneath its orbit, so for observers not at the equator, the satellite never appears directly overhead. Instead, its maximum elevation angle depends on the observer's latitude, the altitude of the satellite orbit when it has the same longitude as the observer, and the altitude of the orbit at apogee and perigee (see the Appendix to Section 5). For example, for an observer at latitude 45° and a satellite in a circular equatorial orbit with an altitude of 500 km, the maximum elevation angle will be only 17°. The maximum elevation angle increases with satellite altitude: for a satellite in geostationary orbit at an altitude of 36,000 km, the highest elevation angle seen by the observer at 45° latitude is 38°, which occurs when the satellite and the observer are at the same longitude.

Because the Earth rotates, a satellite regularly passes directly over parts of the Earth with latitude equal to or smaller than the angle of the satellite's inclination. For areas of the Earth at higher latitudes, the satellite may be observable but it will never be overhead; its maximum elevation angle will be less than 90°. So a satellite in an orbit with an inclination angle of 45° will have a maximum elevation angle of 90° for all points of the Earth between 45° north and 45° south. A satellite in a polar orbit passes directly overhead all points on Earth.

Two types of satellites pass directly over the same area of the Earth on each repetition of their orbit: a satellite in an equatorial orbit at any altitude and a satellite in a geosynchronous orbit. At other points on the Earth, these satellites never appear directly overhead.

The elevation angle of a satellite at a given location has a strong impact on how it can be used, and thus can suggest its purpose. For example, a ground station has a difficult time receiving signals from a satellite that is at a low elevation, for two reasons. First, a signal from the satellite travels a longer distance through dense atmosphere than it would if sent from a satellite at a higher elevation angle, which results in a greater attenuation of the signal strength. Second, objects on the horizon—such as tall buildings or mountains— may be in the line of sight between the ground station and satellite, thereby blocking the signal transmission. In densely built cities, tall buildings can block signals sent to and from a station on the ground for low elevation angles, in the worst cases even up to 70°, so many satellite receivers and transmitters in cities are mounted on the tops of buildings. For some applications, however, the receivers need to be mobile and on the ground, such as ground vehicles using the Navstar Global Positioning System (GPS). GPS satellites operate at 55° inclination and may not be able to adequately serve all their potential users in urban settings. Japan, for example, lies at 30° to 45° latitude so the GPS satellites never pass overhead, and it has high buildings in its urban areas and high mountains in its rural areas that can block GPS signals. It is developing a system of three satellites in highly elliptical orbits with their apogees over Asia (Quazi-Zenith Satellite System (QZSS)). These will work with the US GPS system to better serve Japan's population.

Geostationary communications satellites are also less useful to Russia than to the United States: equatorial orbits do not afford good coverage of the poles or regions at very high latitudes, and many key Russian military installations are in the north polar region. Instead, Russia uses satellites in highly inclined orbits that can be easily seen from northern latitudes during part of their orbit. When these orbits are highly elliptical and have their apogee near the North Pole, the satellites in these orbits appear overhead for longer periods of time, making them particularly useful. Orbits of this type include Molniya orbits, which are discussed below.

Observable Ground Area

How much of the Earth's surface can be observed from a satellite depends on its altitude h. While the altitude determines the maximum area the satellite can observe, the actual observable area may be limited by the sensors the satellite carries, which may not be able to view this entire area simultaneously.

The outer edge of the observable region is a circle, the radius of which depends only on the satellite's altitude. The relation between these two parameters is given in the Appendix to Section 5, and illustrated in Figure 5.4. However, a ground station can generally communicate with a satellite only if it is at an elevation angle greater than some minimum value; this minimum value is typically 5° to 10°. As a result, the effective ground area with which the satellite can communicate is less than the full area the satellite can observe. The radius of the effective region is a function of both satellite altitude h and the minimum elevation angle, as discussed in the Appendix to Section 5.

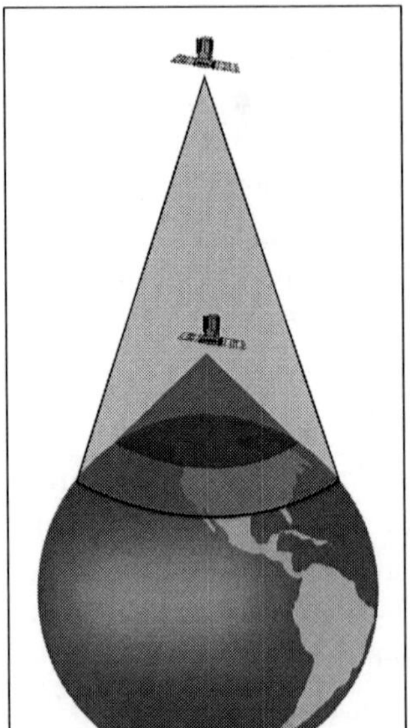

Figure 5.4. The size of the area of the Earth observable from a satellite depends on its orbital altitude. The observable area is compared here for satellites at two different altitudes: the satellite at the lower altitude sees a much smaller area than the one at the higher altitude. Note that the observable area also describes the area on the Earth that can see the satellite.

Table 5.1 lists the radius of the maximum observable ground area by satellites at several altitudes, and the radius of the effective ground area when the minimum elevation angle is 10°. Note that for satellites at low altitudes the effective area is roughly half of the maximum observable area, while for higher orbits the effective area is not reduced as much relative to the maximum area. Note also the much larger fraction of the Earth that is visible from a satellite in geosynchronous orbit compared with low earth orbit.

Table 5.1. This table shows the radius (as measured along the Earth's surface) of the maximum region of the Earth that satellites at several altitudes can see, as well as the percentage of the Earth's surface that region covers. It also shows the size of the effective observable area if the minimum elevation angle at which the ground station can communicate with the satellite is 10°.

	Maximum observable region		Effective observable region (Minimum elevation angle = 10 degrees)	
Satellite altitude (kilometers)	Radius (km)	% of Earth's Surface	Radius (km)	% of Earth's Surface
500	2440	3.6	1560	1.5
1,000	3360	6.8	2440	3.6
20,000 (Semisynchronous)	8450	38	7360	30
36,000 (Geosynchronous)	9040	42	7950	34

Clearly, the higher the orbit, the larger the ground area that the satellite can observe and communicate with. However, other factors also affect the choice of satellite altitude.

The intensity of electromagnetic radiation—including visible light, infrared, and radio waves—drops off in proportion to the square of the distance between the sender and receiver. This drop in signal intensity with increasing altitude suggests that a lower orbit is preferable. On the other hand, for a given coverage area on the ground, a lower orbit satellite must propagate the signal over a wider angle than a higher orbit satellite would (see Figure 5.4). To achieve this wide dispersion, the satellite antenna sacrifices signal strength (or gain) in any single direction, and this partially offsets the distance advantage. Thus, for a given satellite mission, the tradeoffs lie between a low-altitude satellite with a wide-area antenna and a high-altitude satellite with a highly directional antenna.

The satellite's observable ground area, as well as its motion with respect to the Earth, has important consequences for how it can be used. For example, a satellite taking high-resolution photographs of the ground is best placed in a low-altitude orbit. In this case, it will spend only a small fraction of its time in view of any particular location on the Earth. Constant low-altitude surveillance of a specific area would therefore require multiple satellites in orbit, so that one moves into position as another moves out of position.

The total number of satellites required in a constellation (i.e., a system of more than one satellite) in order to have one satellite in the right place at all times is the absentee ratio, which depends on the ground area that each satellite covers. All else being equal, placing the satellites at higher altitudes decreases the absentee ratio. However, using satellites at higher altitudes may be impossible for some applications, such as high-resolution surveillance or ballistic missile defense, which requires space-based interceptors to be relatively close to attacking missiles.

Transmission Time

The round-trip transmission time between a ground station and a satellite is the distance traveled divided by the speed of light (300,000 km/sec). The exact distance traveled depends on the elevation angle of the satellite and where it is in its orbit, but this is roughly twice the orbital height h of the satellite. The round-trip transmission time in seconds is roughly $(2 \times h)/300{,}000$, where h is expressed in kilometers.

For a satellite in geosynchronous orbit at an altitude of 36,000 km, the round-trip transmission time is roughly 0.25 seconds. Because of this time delay, using such a satellite to relay data between two or more ground stations in its field of view requires echo control on telephone transmissions and special protocols for data transmission. In contrast, a satellite at an altitude of 500 km has a round-trip transmission time of only 0.003 seconds, eliminating the need for echo control or other special treatment.

Effects of the Local Environment

Interference

If neighboring satellites use the same transmission frequencies, a receiver on the ground may find that their transmitted signals interfere with one another if the satellites are too close together. Because geostationary orbits are a limited resource, satellites in this orbit are positioned close together and their transmission frequencies must be planned carefully so that their transmitted signals do not interfere with each other. The International Telecommunications Union (ITU) performs the task of assigning locations and frequencies to satellites in geostationary orbit. This is facilitated by the fact that these satellites remain in fixed positions with respect to each other. When this arrangement is not respected or the orbit becomes too crowded, interference problems have occurred.[3] Satellites in different orbits are not stationary with respect to each other, so they pose a more difficult interference problem if they communicate at the same frequencies.[4] Recently, the ITU designed regulations that allow satellites in low earth orbits to use the same frequencies as

---

3. In 1996, the United Nations reported that severe crowding in the geostationary orbital slots over Asia "led to the jamming of a communication satellite by PT Pasifik Satellite Nusantara (PSN) of Jakarta, Indonesia, in defense of an orbital position claimed by Indonesia. This incident focused global attention on the worsening problem of orbital crowding and caused the matter to be brought before the 1997 World Radio Communication Conference (WRC) of the 187 member-nation ITU in Geneva" (United Nations, Highlights in Space: Progress in Space Science, Technology, Applications, International Cooperation and Space Law, 1997 [Vienna: United Nations, 1998], 51).
4. The Skybridge low-earth-orbiting satellite system (www.skybridgesatellite.com, accessed January 15, 2005) operates in the Ku band, as do many geostationary satellites. To avoid interference, the Skybridge system turns off its transponders as the satellite approaches the equatorial plane where the geostationary satellites are. Skybridge needs to have more than one satellite in view at any one time to avoid interruption of service, adding complexity and expense to the system. See Kristi Coale, "Small Satellites Push for Elbow Room," Wired News, October 14, 1997, http://www.wired.com/news/technology/0,1282,7657,00. html?tw=wn story _mailer, accessed January 15, 2005.

those in geostationary orbits by setting an upper limit to the amount of power they may broadcast.[5]

Radiation Environment

Space is a harsh environment.[6] Satellites do not enjoy the protection of the atmosphere against radiation and particles from the Sun and the larger universe. This radiation environment and its changes are sometimes referred to as space weather.

Solar ultraviolet and X-rays generally do not penetrate the skin of a satellite. Only during solar flares do X-rays have sufficient energy to penetrate a few millimeters of aluminum, but generally the frequency and duration of this radiation is not sufficient to damage internal electronics. These X-rays do, however, damage and degrade solar panels, which many satellites rely on to generate energy.

Charged particles (positively charged protons and negatively charged electrons) are an additional concern. These particles can have very high energies and can damage and degrade electronics. These particles are trapped by the Earth's magnetic field, forming two toroidal (doughnut-shaped) regions around the Earth's equator known as the Van Allen belts (see Figure 5.5).

The inner torus extends from an altitude of roughly 500 km to 5,500 km, with the highest particle density in the middle at about 3,000 km.[7] The particle density is greatest at the equator and low latitudes, then decreases as the latitude increases. By latitude 50° or 60° north or south, the density in the belt is very low. This inner belt is populated primarily with high-energy protons that can readily penetrate spacecraft and, with prolonged exposure, damage instruments and endanger astronauts. (This region also contains highenergy electrons, but at a much lower density than the protons.) Both manned and unmanned spaceflights stay out of the high-density regions.

5. The ITU amended the radio regulations at the World Radiocommunication Conference in 2000. The documentation pertaining to the regulation can be obtained from the ITU website http://www.itu.int, accessed January 15, 2005.
6. See "The Earth's Trapped Radiation Belts," NASA Space Vehicle Design Criteria (Environment), NASA SP-8116, (Springfield, Virginia: March 1975), available from the National Technical Information Service; NASA JPL Radiation Effects Group, "Space Radiation Effects on Microelectronics," http://parts.jpl.nasa.gov/docs/Radcrs_Final.pdf, accessed January 15, 2005.
7. The belts will reach down to lower altitudes near the poles, as shown in Figure 5.5, but the particle density is low at that point.

The outer torus extends from an altitude of roughly 12,000 to 22,000 km, with the highest particle density also in the middle, at 15,000 to 20,000 km. The outer belt is populated primarily with high-energy electrons. The electrondistribution is much more variable spatially and temporally than the proton distribution. However, like the proton flux, the electron flux is highest near the equator and becomes negligible at a latitude of 60° north or south.

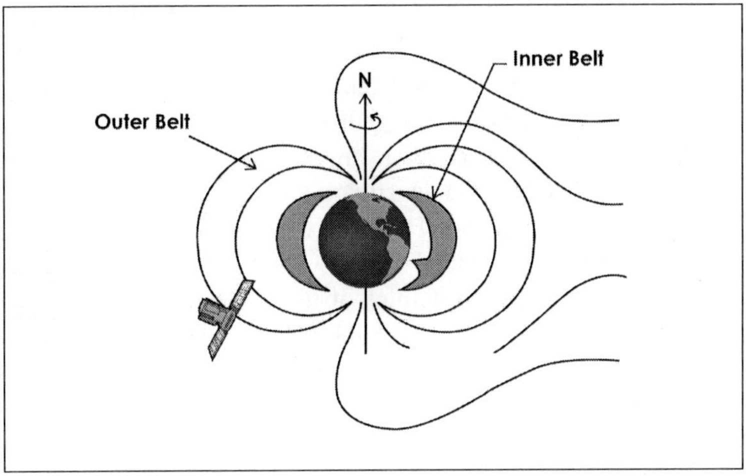

Figure 5.5. An illustration of the Van Allen radiation belts,[8] showing the inner radiation belts, which have a maximum density at an altitude of about 3,000 km and are primarily protons, and the outer belts, which have a maximum density at about 17,000 km and are primarily electrons. The solar wind (consisting mainly of low energy protons and electrons) distorts the toroidal shape and flattens the torus on the sunward side and creates a tail on the shaded side, which is shown by the solid lines on the right.

A region between the inner and outer belts, known as the "slot," has a low density of high-energy electrons. This region extends from roughly 6,000 to 12,000 km, but can disappear during active solar periods when the inner and outer belts sometimes overlap.

Because the radiation belts are well characterized, they can be accounted for in satellite design. But radiation shielding adds mass

8. Adapted from NASA JPL Radiation Effects Group, "Space Radiation Effects on Microelectronics," presentation, 2002, http://parts.jpl.nasa.gov/docs/Radcrs_Final.pdf, accessed January 15, 2005.

and expense. A few centimeters of aluminum are sufficient to stop most electrons. However, as the electrons entering the aluminum slow down, they can generate X-rays. Thus satellites in the most intense radiation environments may require heavy lead shielding to protect against these X-rays. In addition, extraordinary events can significantly change the radiation environment and endanger satellites not designed to withstand higher fluxes. Five or ten times a year, solar flares generate bursts of high-energy protons and electrons, which may occasionally be stronger than expected.

Thus the choice of orbit determines the amount of shielding a satellite needs, based on whether and what part of the Van Allen belts they will encounter. For example, at altitudes of less than 500 to 1,000 km, the density of charged particles is low and little shielding is required. These are the altitudes at which many satellites, and all extended missions with personnel, operate. On the other hand, the satellites of the U.S. Global Positioning System operate at an altitude of 20,000 km—where the density of highenergy electrons is at or near its maximum. This demonstrates the feasibility of shielding against even high concentrations of charged particles.

Human actions can alter the radiation environment as well. In 1962, the United States conducted a nuclear test explosion called Starfish at an altitude of 400 km. It generated a significant perturbation in the trapped electron distribution. The maximum intensity of the perturbation was in the region over the equator at altitudes between 1,600 and 6,300 km, but effects reached out to at least 44,000 km. Below 1,600 km, the perturbation decayed fairly rapidly, but at these altitudes, the radiation level was still an order of magnitude higher than normal several years after the test.

Effects of the Atmosphere

There is no outer "edge" to the atmosphere. The air that makes up the atmosphere is held to the Earth by gravity, just as the water in the oceans is. And just as the water pressure increases with depth in the ocean, the atmosphere is most dense at ground level and thins out quickly with increasing altitude, falling off roughly exponentially. At 10 km altitude (the height of Mount Everest) the air is nearly too thin to breathe, and the density is about onethird of the density at sea level. At 100 km altitude, the density has dropped to less than one-millionth of that at sea level; by 600 km it is reduced by another factor of one million. For many purposes, the "sensible atmosphere" ends around 100 km, and this is generally accepted as the altitude at which "space" begins. However, for some purposes the effects of the atmosphere must be considered at altitudes higher than 100 km. For example, the atmospheric drag on satellites may be very small at altitudes of several hundred kilometers, but its cumulative effect over many orbits is not negligible.

The atmosphere has important consequences for satellites in low orbits. One consequence of the balance of forces (centrifugal and gravity) that keep a satellite in orbit is that if atmospheric drag begins to slow a low-orbit satellite, it will no longer be moving fast enough to stay in its original orbit and will begin to spiral down toward the Earth. The increase of atmospheric drag at lower altitudes restricts satellite orbits to altitudes of a few hundred kilometers or higher. Satellites in low orbits must carry stationkeeping fuel so they can occasionally maneuver to offset the effects of atmospheric drag and stay in orbit.

Moreover, because the atmospheric density varies spatially and temporally, there is an inherent limit to the accuracy with which the future position of a satellite in low orbit can be predicted. This fact is important for long-term monitoring of satellites.

More Complicated Effects of Gravity

The circular and elliptical orbits described in Section 4 considered only two forces on the satellite: Earth's gravity and centrifugal force. In these cases, Earth's gravity is assumed to be spherically symmetrical. However, because the Earth is not a perfect sphere, its gravitational field is not perfectly symmetric and this affects the orbital shape and orientation—and changes them over time—in subtle but important ways. The gravitational pull of the Sun, which is much weaker than that of the Earth, also affects the orbits.

In the absence of these gravitational irregularities, the orientation of the orbital plane of a satellite would remain constant in space. However, because of them, the orbital plane precesses, that is, it rotates with respect to the Earth's axis. The rate of precession depends on the eccentricity, altitude, and inclination of the orbit. The proper choice of these parameters can allow a satellite to use the precession rate to aid a specific purpose. For example, by choosing an orbit that precesses at a specific rate, the orbital plane can be made to keep a constant angle with respect to a line between the Earth and Sun throughout the year, as the Earth travels around the Sun. A satellite in such an orbit—a sun-synchronous orbit—observes each place on Earth at the same local time and sun angle; more detail on this orbit appears below.

Gravitational irregularities also cause the major axis of an elliptical orbit to rotate slowly in its orbital plane. Thus, for an elliptical orbit inclined with respect to the equator, its apogee moves slowly from over one hemisphere (i.e., northern or southern) to over the other, then returns over the first. The rate at which this occurs depends on the inclination angle of the orbit; for an angle of 63.4°, the rate is zero. Thus, the apogee of an orbit with this inclination angle remains over the same hemisphere. In fact, while precession still causes the apogee to slowly rotate about the Earth's axis, it remains over the same latitude as it rotates.

These and other orbits are discussed in more detail below.

Common Orbits

Low Earth Orbits

Satellites in low earth orbits (LEO) operate at altitudes of hundreds of kilometers up to around 1,000 km. (Satellites at orbital heights of a few thousand kilometers could also be said to be in low earth orbits, but few satellites populate this part of space because of the large amount of radiation there.) LEO satellites have orbital periods of roughly 90 minutes. As noted above, space at these altitudes is mostly free from high radiation and charged particles. Atmospheric drag is small above a few hundred kilometers, although increases in solar activity can cause the outer layer of the Earth's atmosphere to expand, increasing the drag on satellites in orbits in the lower part of this region.

Since a satellite in LEO cannot see a large ground area and since it moves relative to the Earth's surface, LEO may not seem to be useful for missions such as communications. However, a network that contains enough LEO satellites to see all regions of the Earth and that can relay signals between the satellites can provide continuous worldwide coverage. If such a network includes polar or near-polar orbits, it can also provide coverage of polar and high latitude regions, as geostationary satellites cannot. Because they are in low orbits, the round-trip transmission time from these satellites is relatively short (0.005 seconds to and from the ground), eliminating the need for echo control or other special treatment. (The time required for signals transmitted over long distances around the Earth when relayed through multiple satellites is dominated by the distance along the Earth rather than the altitude of the satellite: transmission halfway around the Earth—20,000 km—requires at least 0.067 seconds.[9]) Moreover, if some of the satellites are on highly inclined orbits, observers at high latitudes can see the satellites at high elevation angles, which reduces interference with the signals by buildings and other objects. These qualities make LEO orbits useful for personal communications systems.

The disadvantage of using LEO satellites for this purpose is that the network requires many satellites. Recall that any observer sees a satellite passing overhead for roughly 10 minutes out of its 90-minute orbit, so nine satellites would be required to provide continuous coverage of a single band on the Earth around its ground track (for an orbital altitude of 500 km, the width of this band is roughly 3,000 km, as Table 5.1 shows). For broader coverage, considerably more satellites would be needed. For example, the Iridium constellation, which is used for a variety of military and commercial purposes, has 66 satellites distributed in six different orbits with an altitude of 780 km. The six orbits are in six different orbital planes, each at an inclination angle of 86.4°.[10]

Low earth orbits may also be useful for missions that do not require realtime communication. Such missions may need only one or a few satellites. For example, data need not be sent to ground users immediately, but can be stored and then forwarded

9. The relay process, which requires each relay satellite to receive and retransmit the signal, also adds to the transmission time.
10. "Iridium Satellite Solutions," http://www.iridium.com, accessed January 15, 2005.

when the satellite passes over the ground station (this arrangement is known as "store-and-forward").

For missions that are not time critical, the motion of the LEO satellites relative to the Earth means that a single satellite in polar orbit can cover the entire Earth. If the orbital period is chosen so that the ground coverage areas on successive orbits lie next to each other, a satellite in a polar orbit can see any spot on Earth twice a day.

Some missions require low orbits. Earth observation and reconnaissance satellites intended to take high-resolution images of the Earth must be close to the Earth to get such resolution (see discussion in Appendix B to Section 11). For example, the U.S. Keyhole satellites, which took optical photographs for intelligence purposes, were usually deployed in elliptical orbits with an apogee and perigee at 1,000 and 300 km, respectively.[11] These have been replaced by a new generation of imaging satellites in similar orbits. Since these satellites move with respect to the Earth, they cannot offer continuous coverage of a particular area.

Circular Medium Earth Orbits

Satellites in circular medium earth orbits (MEO), also termed intermediate circular orbits (ICO), have altitudes between those of low earth orbits and geosynchronous orbits: from roughly 1,500 to 36,000 km.

A common orbit is one with an altitude of roughly 10,000 km and an orbital period of about 6 hours. Continuous worldwide real-time coverage can be obtained with fewer satellites than are needed for a constellation of satellites in low earth orbits. For example, the ICO communications satellite system under construction will consist of 10 satellites in 2 orbits at an altitude of 10,390 km. The two orbital planes will be at 45° inclination, rotated 180° around the Earth's axis with respect to one another.[12]

Satellites in such medium earth orbits are relatively slow moving as seen from the Earth, thus requiring fewer and simpler

11. Federation of American Scientists, "KH-11 Kennan/Crystal Satellites," September 9, 2000, http://www.fas.org/spp/military/program/imint/kh-11.htm, accessed January 15, 2005.
12. "ICO Satellite Wireless Services," http://www.ico.com, accessed January 15, 2005, and "Lloyd's Satellite Constellations," http://www.ee.surrey.ac.uk/Personal/L.Wood/ constellations/ ico.html, accessed January 15, 2005.

handover arrangements than a LEO system. The round-trip transmission time to these satellites from the ground is longer than to a satellite in low earth orbit: the ICO transmission time is 0.069 seconds, whereas for the Iridium system it is 0.0052 seconds. This longer transmission time is less of an issue for communications over long distances (a signal traveling halfway around the world would along the Earth's surface require a minimum of 0.067 seconds, comparable to the time it takes for a round trip to the ICO satellite), and using higher altitude satellites reduces the number of satellites the signal must be relayed between to cover long distances. However, satellites in MEO orbits must employ radiation- hardened components (particularly to protect their computer systems) to survive long term.

A special type of medium earth orbit is the semisynchronous orbit, which has a period of 12 hours and an altitude of roughly 20,000 km. Both the U.S. Navstar Global Positioning System (GPS) and Russian Glonass navigational satellites use these orbits. A navigational system needs at least four satellites within view of the user at all times, where a continuous communications system needs only one. Thus, a navigational system requires more satellites than does a communications system deployed at the same altitude: both GPS and Glonass (when fully deployed) use 24 satellites. The GPS satellites are in six orbital planes at an inclination angle of 55°; Glonass is designed to use three orbital planes at an inclination angle of 65°.

Molniya Orbits

Molniya orbits are highly elliptical, with a period of 12 hours and an inclination of 63.4°. At this inclination, the apogee remains over the same latitude in the northern (or southern) hemisphere, rather than precessing. The Soviet Union first used this type of orbit for its Molniya satellite system, hence the name. They are sometimes referred to as highly elliptical orbits (HEO).

A satellite in a highly elliptical orbit with the apogee over the northern hemisphere covers Earth's high-latitude regions for a large fraction of its orbital period. As discussed in Section 4, the speed of a satellite is not constant on an elliptical orbit. The satellite has a high speed as it traverses the orbit near perigee and moves slowly near apogee—thus spending most of its time in the sky over the northern hemisphere.

The Russian Molniya satellites are in orbits with an apogee of roughly 40,000 km and a perigee of roughly 1,000 km (or an eccentricity of 0.75). For eight of their 12-hour periods, each satellite remains visible to the regions under the apogee, with elevation angles above 70°. A constellation of three satellites, with their major axes oriented at 120° with respect to each other, ensures continuous coverage of this area.

Molniya orbits are also used by U.S. intelligence satellites that monitor Russia and by Russian early warning satellites that watch for U.S. missile launches.

Tundra Orbits

Like Molniya orbits, Tundra orbits have an inclination of 63.4°, so their apogees remain over one hemisphere. They are typically used to provide coverage of high latitude areas, with their apogee over the northern hemisphere. However, they are not as highly elliptical as are the Molniya orbits, and their period is 24 hours rather than 12.

Satellites in Tundra orbits are visible to the regions under the apogee for 12 of their 24-hour periods. Thus, it is possible to obtain continuous coverage of this region with only two satellites whose orbits are rotated 180° with respect to each other. The Russian Tundra system uses two satellites in orbits with an apogee and perigee of roughly 54,000 and 18,000 km, respectively.

Geostationary Orbits

Geosynchronous orbits have a period equal to the Earth's rotation period. The most useful geosynchronous orbit is the geostationary orbit, which is a circular orbit at an altitude of 35,786 km in the equatorial plane. Because a geostationary satellite appears as a fixed point in the sky to all observers on the ground, users need no tracking equipment to send or receive signals from the satellite. Three satellites can provide worldwide coverage, excluding the polar regions. The area of visibility of the satellite is large; it is not quite half the Earth—about 43% coverage. Thus, geostationary satellites can provide continuous service over a wide geographical area. This is very useful for television and radio broadcasting, since it permits real-time data transfer over a wide geographic area without using a store-and-forward scheme. It also provides the necessary flexibility for commercial and military communications, which need to support users from widely different, nonpredetermined locations.

Geostationary satellites operate outside the densest regions of the Van Allen belt, but they are subject to infrequent bursts of high-energy particles from the Sun that can damage or degrade them.

Sun-Synchronous Orbits

Satellites in sun-synchronous orbits pass over a given part of the Earth at roughly the same local time of day (though not necessarily every day). That is, whenever the satellite observes a given ground location, the Sun is always in the same location in the sky. Such orbits are particularly useful for missions that take images of the Earth, because shadows from objects at a given location on the Earth's surface are always cast from the same angle. This simplifies the comparison of images taken on different days to detect changes. Satellites in these orbits are often placed at low altitudes (with short periods) so that they provide complete coverage of the Earth's surface at least once per day.

The inclination of sun-synchronous orbits is chosen so that the precession of the orbital plane around the Earth due to gravitational irregularities keeps the plane at a constant angle with respect to a line between the Earth and Sun throughout the year. The precise inclination that produces this effect depends on the orbit's altitude and eccentricity; it is typically 96–98°, making the orbits slightly retrograde. Figure 5.6 illustrates how a nonprecessing orbit differs from an orbit that precesses synchronously with the Sun.

In a special sun-synchronous orbit, called a dawn-to-dusk orbit, the satellite's orbital plane coincides with the plane that divides the half of the Earth that is illuminated by the Sun from the half that is dark. If the plane were aligned slightly differently, the satellite would spend half of its time in full sunlight and half in shadow, but a dawn-to-dusk orbit allows the satellite to always have its solar panels illuminated by the Sun. For example, the Canadian

Radarsat Earth observation satellites13 use such a dawn-to-dusk orbit to keep their solar panels facing the Sun almost constantly, so they can rely primarily on solar power and not on batteries.

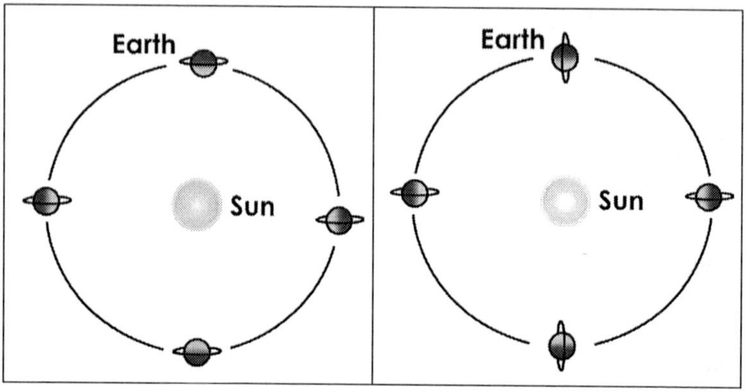

Figure 5.6. Both panels show the Earth at four positions in its yearly orbit around the Sun, and the orbital plane of the same satellite in each case. Panel A shows a case in which the satellite's orbit does not precess and remains in a fixed orientation with respect to space. Thus, a satellite that is directly above a location on the Earth when the local time is midnight and noon, would four months later observe this location when the local time is 6 am and 6 pm. Panel B shows a sun-synchronous orbit. The orbit is in a plane chosen to precess at a rate synchronized with the Earth's trip around the Sun, so that the plane maintains a constant angle throughout the year with respect to a line between the Earth and Sun. As a result, during the entire year, this satellite observes a point on the Earth at the same local time.

In a special sun-synchronous orbit, called a dawn-to-dusk orbit, the satellite's orbital plane coincides with the plane that divides the half of the Earth that is illuminated by the Sun from the half that is dark. If the plane were aligned slightly differently, the satellite would spend half of its time in full sunlight and half in shadow, but a dawn-to-dusk orbit allows the satellite to always have its solar panels illuminated by the Sun. For example, the Canadian Radarsat Earth observation satellites[13] use such a dawn-to-dusk orbit to keep their solar panels facing the Sun

13. For information on Radarsat, see the Radarsat International website http://www.radarsat2.info/, accessed January 15, 2005.

almost constantly, so they can rely primarily on solar power and not on batteries.

Lagrange Points

There are five special orbits in which satellites orbit not the Earth but the Sun, and do so in a way that they maintain a fixed position relative to the Earth as it orbits the Sun. These fixed locations are called Lagrange points; there are five such points, one corresponding to each of the five orbits (see Figures 5.7 and 5.8).

A satellite orbiting the Sun closer than the Earth does has a shorter orbital period than the Earth's. However, such a satellite is pulled by the Earth's gravitational field as well as by that of the Sun. This effect is negligible if the satellite is far from the Earth, but must be taken into account for a satellite close to Earth. For a satellite directly between the Earth and Sun, the direction of the Earth's pull is exactly opposite that from the Sun, effectively canceling some of the Sun's gravitational pull. At the first Lagrange point[14] (L_1), the net gravitational force on the satellite is the same as the Sun's gravitational force on the Earth, so that the satellite orbits the Sun with the same orbital period as the Earth. A satellite in this position stays with the Earth throughout its journey around the Sun. The L_1 point is about four times more distant from the Earth than the Moon is. The L_1 point is particularly useful for scientific missions that study the Sun, and satellites positioned there can give early warning of increased solar winds.

A second Lagrange point is located the same distance from the Earth but on the other side, directly away from the Sun. In this case, the Earth's gravitational pull adds to that of the Sun, increasing the orbital speed required for the satellite to stay in orbit. In this case, the satellite keeps up with the Earth in its orbit, while it would normally fall behind. Scientific missions are positioned there as well, allowing the satellite to be maximally far from the Earth (to minimize interference), but maintain constant contact. NASA plans to place its Next Generation Space Telescope (NGST), the successor to the orbiting Hubble telescope, at or near L_2.

14. There are analogous Lagrange points for the Earth-Moon system. These points are near the Moon and stationary with respect to it. The Lagrange points discussed here are all in the Earth-Sun system.

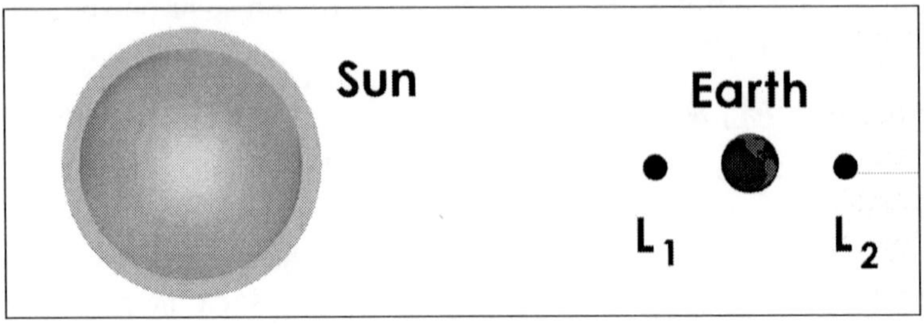

Figure 5.7. The L_1 and L_2 Lagrange points for the Earth-Sun system.

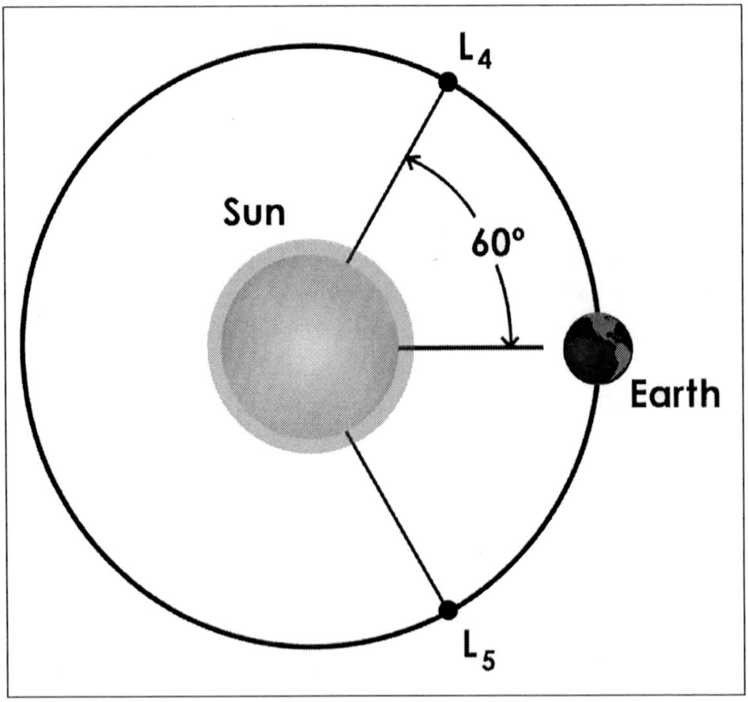

Figure 5.8. The L_4 and L_5 Lagrange points for the Earth-Sun system.

The L_3 point, which lies on the other side of the Sun, directly opposite the Earth, is not very useful for satellites. The L_4 and L_5 points are along the Earth's orbit, but precede and lag it. They are 60° away from the Earth-Sun line. Some suggest the L_2 will be strategically interesting for space exploration or for space militarization.[15] Since craft at L_2 are in a stable position and need

15. For example, see James Oberg, "Will China's Space Plan Skip the Moon?" Space News, May 24, 2004, 13.

little fuel to remain there for extended periods of time, L_2 could be used as a place to assemble other spacecraft from parts lifted piece by piece. Such a scheme could be more energy efficient than trying to assemble large structures on the Moon and more feasible than assembling them on Earth and then launching them. Objects at L_2 are also out of easy observation by the Earth, being quite distant.

Appendix - 5

Details of Elevation Angle and Ground Area

Elevat Ion Angle

The elevation angle ε of a satellite as seen by an observer is[1]

$$\varepsilon = \arctan\left[\frac{\cos\phi - R_e/(R_e+h)}{\sin\phi}\right] \quad (5.1)$$

where R_e is the radius of the Earth, h is the altitude of the satellite, and φ is defined as

$$\cos\phi = \cos(\psi - \lambda)\cos l \cos\phi + \sin l \sin\phi \quad (5.2)$$

where l and ψ are the latitude and longitude of the observer, respectively, and φ and λ are the latitude and longitude of the satellite.

For a satellite in an equatorial circular orbit, φ = 0. The maximum elevation angle ε_{max} seen by the observer occurs at the point in the orbit for which the longitude of the observer equals that of the satellite (ψ = λ), so the above equation simplifies to

$$\cos\phi = \cos l \quad (5.3)$$

Thus, the maximum elevation angle ε_{max} of a satellite in an equatorial circular orbit as seen by an observer at latitude l is

$$\varepsilon_{max} = \arctan\left[\frac{\cos l - R_e/(R_e+h)}{\sin l}\right] \quad (5.4)$$

1. Gérard Maral and Michel Bousquet, Satellite Communications Systems, 4th ed. (West Sussex, England: Wiley, 2002), 44-45.

For an observer at a latitude of 45°, ε_{max} is 17° for h = 500 km and 38° for h = 36,000 km. Recall that tall buildings can interfere with satellite reception for elevation angles of up to 70°. Using this equation, for a satellite in geostationary orbit (with h = 36,000 km) this corresponds to an observer at a latitude of roughly 18°.

Ground Area Viewed by a Satellite

From Figure 5.9, it can be shown that the radius R_{area} of the maximum circular region (as measured along the Earth's surface) that can be viewed by a satellite at altitude h is equal to

$$R_{area} = R_e \cos^{-1}\left(\frac{R_e}{R_e + h}\right) \qquad (5.5)$$

where R_e is the radius of the Earth and the angle is in radians. The fraction F of the Earth's surface this region represents is

$$F = 0.5\,(1 - \cos(R_{area}/R_e)) = 0.5h/(R_e + h) \qquad (5.6)$$

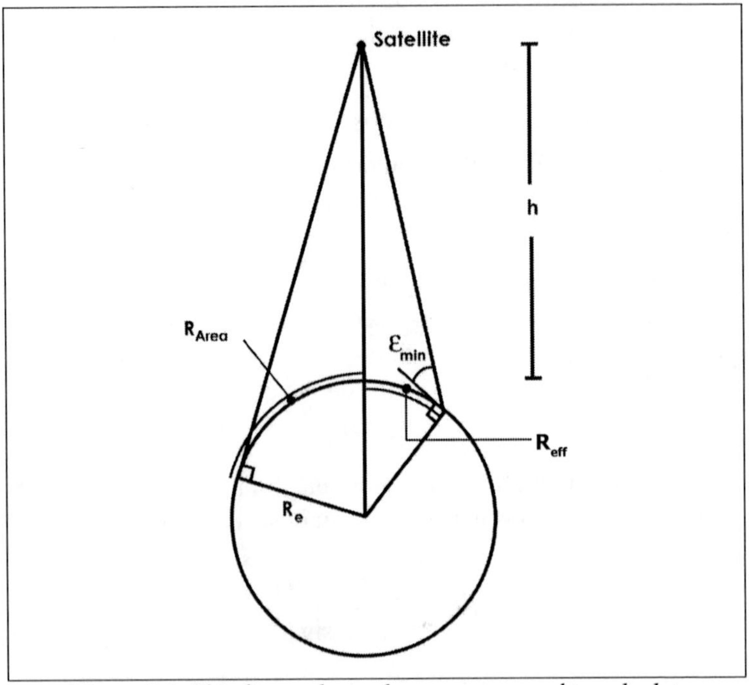

Figure 5.9. This figure shows the geometry used to calculate the ground area visible to a satellite at altitude h.

If the minimum elevation angle ε_{min} at which the user can communicate with the satellite is greater than 0°, then by using the law of sines, it can also be shown, from Figure 5.9, that the radius R_{eff} of the effective observable region (as measured along the Earth's surface) is

$$R_{eff} = R_e \left(\frac{\pi}{2} - \varepsilon_{min} - \sin^{-1}\left(\frac{R_e \cos \varepsilon_{min}}{R_e + h} \right) \right) \tag{5.7}$$

where the angles are expressed in radians. The fraction of the Earth's surface covered by this region is

$$F = \frac{1}{2}\left[1 - \cos\left(\frac{R_{eff}}{R_e} \right) \right]$$

$$= \frac{1}{2}\left[1 - \sqrt{1 - x^2}\,\sin \varepsilon_{min} + X \cos \varepsilon_{min} \right] \tag{5.8}$$

where

$$X = \frac{R_e \cos \varepsilon_{min}}{R_e + h} \tag{5.9}$$

Section - 6

Maneuvering in Space

To maneuver, a satellite in orbit must use rocket engines (thrusters) to change the magnitude or direction of its velocity. Because the orbital speed of satellites is so large, the velocity changes required for maneuvering may also be large, requiring the thrusters to use large amounts of propellant.

How much and how quickly a satellite can maneuver depends on the amount and type of propellant it carries. There are practical limits to the amount of propellant a satellite can carry since it increases the total mass that must be launched into orbit. These constraints on maneuvering in space have important consequences for satellite operations.

This section discusses the different types of satellite maneuvers and the changes in satellite velocity required for each. Section 7 outlines the amount of propellant required for these maneuvers.

Basic Satellite Maneuvers

When a satellite maneuvers, it changes orbit. Since the speed of a satellite is related to its orbit, maneuvering can be complicated. Three basic maneuvers are used to change orbits: (1) changing the shape or size of an orbit within the orbital plane; (2) changing the orbital plane by changing the inclination of the orbit; and (3) changing the orbital plane by rotating the plane around the Earth's axis at constant inclination. (Recall that all satellite orbits lie in a plane that passes through the center of the Earth.) We discuss each of these in more detail below, as well as several common orbital changes that use these basic maneuvers. Maneuvers within the orbital plane allow the user to change the altitude of a satellite in a circular orbit, change the shape of the orbit, change the orbital period, change the relative location of two satellites in the same orbit, and de-orbit a satellite to allow it to return to Earth. To indicate the scale of velocity changes required for some common orbital maneuvers, Table 6.1 lists such maneuvers along with a characteristic value of the velocity change needed in each case (see the Appendix to Section 6 for more details).[1]

A velocity change is typically referred to as delta-V, or ΔV, since the term "delta" is commonly used in technical discussions to indicate a change in some quantity. To get a feel for what these numbers mean, it is helpful to keep in mind that a speed of 1 km/s is roughly four times faster than a passenger jet. In addition, as Section 7 shows, generating a velocity change of 2 km/s with conventional propulsion technologies would require a satellite to carry its own mass in propellant—thus doubling the mass of the satellite.

Maneuvers that change the orbital plane of a satellite can require very large changes in the satellite's velocity, especially for satellites in low earth orbit (see Table 6.1). This has important implications for the feasibility and utility of space-based systems that require such maneuvers.

Table 6.1. This table shows the change in satellite velocity (ΔV) required for various types of maneuvers and activities in space, where $\Delta\theta$ is the change in inclination.

Type of Satellite Maneuver	Required ΔV (km/s)
Changing orbital altitude within LEO (from 400 to 1,000 km)	0.3
Stationkeeping in GEO over 10 years	0.5–1
De-orbiting from LEO to Earth	0.5–2
Changing inclination of orbital plane in GEO	
by $\Delta\theta = 30°$	2
by $\Delta\theta = 90°$	4
Changing orbital altitude from LEO to GEO (from 400 to 36,000 km)	4

1. A general maneuver will be combination of these basic maneuvers. Designing a maneuver that changes the altitude and orbital plane at the same time, rather than through sequential maneuvers, can reduce the velocity change required.

Changing inclination of orbital plane in LEO	
by $\Delta\theta = 30°$	4
by $\Delta\theta = 90°$	11

These numbers are calculated in the Appendix to Section 6. (LEO = low earth orbit, GEO = geosynchronous orbit)

Maneuvers Within the Orbital Plane

Maneuvers that change the shape or size of a satellite's orbit without changing its orbital plane can be made by changing the magnitude but not the direction of the velocity. These kinds of maneuvers can require significantly less ΔV than maneuvers that change the direction of the velocity.

Changing the Shape of the Orbit

Consider a satellite that is initially in a circular orbit with altitude h. As discussed in Section 4, the laws of physics require it to have a particular speed for that altitude, which is given by Figure 4.1 and Equation 4.2. If the speed of the satellite is suddenly increased by ΔV at some point on the orbit (without changing the direction of the velocity), the satellite does not go faster around the same orbit; instead, the orbit becomes an ellipse in the same orbital plane (see Figure 6.1). The perigee of the new orbit (where the satellite is closest to Earth) lies at the point where the speed was increased, and this point will remain at an altitude h. As is always the case for elliptical orbits, the major axis passes through the center of the Earth, with the perigee and apogee of the new orbit at opposite ends. The orbital altitude at apogee is greater than h and depends on the value of ΔV, as discussed in the Appendix to Section 6.

If the speed of a satellite on a circular orbit is reduced at some point on the orbit by thrusting in the direction opposite to the satellite motion, that point becomes the apogee of an elliptical orbit, with an altitude of h at apogee. The perigee then lies at an altitude less than h. As shown in the Appendix to Section 6, a relatively small value of ΔV results in a significant change in altitude at apogee. As an example, for a satellite in orbit at an altitude of 400 km, a ΔV of 0.1 km/s would lead to a change in altitude at apogee of 350 km, so that apogee lies at an altitude of

750 km. In the more general case of an elliptical orbit, changing the speed but not the direction of the velocity of the satellite results in another elliptical orbit, but of a different shape and orientation within the plane. The resulting orbit depends on both the value of ΔV and the point at which the velocity changed. However, in two specific cases, an elliptical orbit can be changed into a circular orbit, with one of two altitudes. Increasing the speed at apogee by the required amount results in a circular orbit with an altitude equal to that at the apogee of the ellipse. Decreasing the speed at perigee by a specific amount results in a circular orbit with an altitude equal to that at the perigee of the ellipse.

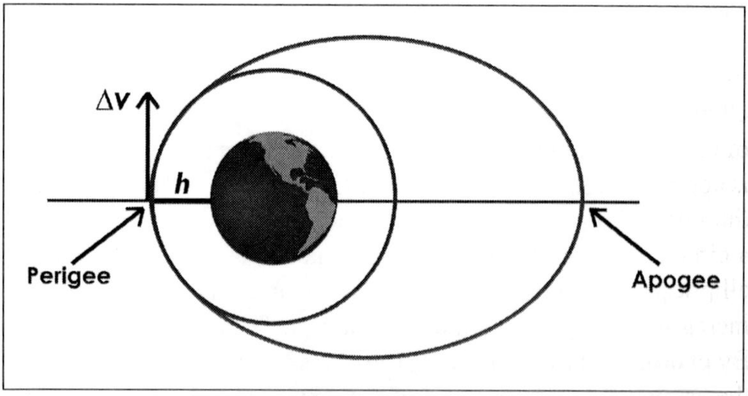

Figure 6.1. The speed of a satellite in a circular orbit with altitude h is increased by Δ V at the point shown. This point becomes the perigee of the new elliptical orbit that the satellite will follow.

If the speed of a satellite on a circular orbit is reduced at some point on the orbit by thrusting in the direction opposite to the satellite motion, that point becomes the apogee of an elliptical orbit, with an altitude of h at apogee. The perigee then lies at an altitude less than h. As shown in the Appendix to Section 6, a relatively small value of ΔV results in a significant change in altitude at apogee. As an example, for a satellite in orbit at an altitude of 400 km, a ΔV of 0.1 km/s would lead to a change in altitude at apogee of 350 km, so that apogee lies at an altitude of 750 km. In the more general case of an elliptical orbit, changing the speed but not the direction of the velocity of the satellite results in another elliptical orbit, but of a different shape and orientation within the plane. The resulting orbit depends on both the value of ΔV and the point at which the velocity changed. However, in two specific cases, an elliptical orbit can be changed into a circular orbit, with one of two altitudes. Increasing the speed at apogee by the required amount results in a circular orbit with an altitude equal to that at the apogee of the ellipse. Decreasing the speed at perigee by a specific amount results in a circular orbit with an altitude equal to that at the perigee of the ellipse.

Changing the Altitude of a Satellite in a Circular Orbit

The strategy described above to change the shape of the orbit, can also be used to increase the altitude of a circular orbit from h1 to h2, through a twostep process (see Figure 6.2). The first step is to increase the speed of the satellite by ΔV1 so that the resulting elliptical orbit has an altitude at apogee of h2. Recall that the perigee of the new orbit lies at the point where the velocity increase (ΔV1) is applied and has an altitude of h1. Once this is done, the speed of the satellite at apogee is less than its speed would be if it were on a circular orbit with altitude h2. The second step is to change the elliptical orbit the satellite is on to a circular one at altitude h2 by increasing the speed at apogee by the appropriate amount (ΔV2). By choosing ΔV1 to make the apogee of the elliptical orbit at h2, the satellite's velocity will be tangent to the larger circular orbit (at point P2 in Figure 6.2), and ΔV2 needs to change only the satellite's speed and not its direction.

The total ΔV required to make the orbital change described above is the sum of the velocity changes applied in each of these two steps: $\Delta V = \Delta V1 + \Delta V2$. These velocity changes are calculated in the Appendix to Section 6. The elliptical orbit used to move between these two circular orbits, which is tangent to both orbits, is called a Hohmann transfer orbit (see Figure 6.2). This method is fuel-efficient since it requires the minimum ΔV needed to transfer between two orbits. The time required for such a transfer is half the period of the elliptical transfer orbit. This time can be shortened and the transfer done more quickly by applying a larger ΔV1 in the first step of the process than that described above. In this case, the velocity of the satellite will not be tangent to the larger circular orbit when it reaches that orbit, so ΔV2 will need to adjust the speed of the satellite as well as rotate its direction to put it on the circular orbit. Since both ΔV1 and ΔV2 will be larger in this case, it is clear that using the Hohmann transfer orbit requires the minimum energy for this transfer. Satellites placed in geostationary orbits are frequently placed in a low earth orbit initially, and then moved to geostationary orbit using a Hohmann transfer orbit.

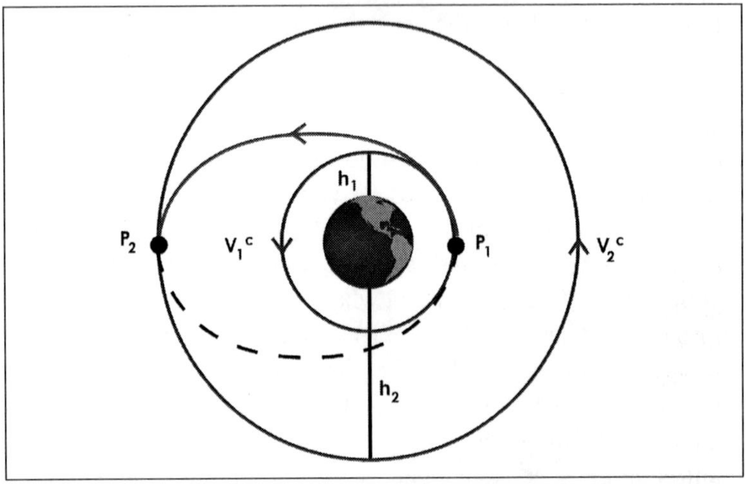

Figure 6.2. This figure shows the elliptical Hohmann transfer orbit between two circular orbits, the initial one with altitude h1 and the final one with altitude h2. ΔV1 is applied at point P1 and ΔV2 at P1. The satellite travels only half an orbit on the transfer orbit; it does not travel the half of the ellipse indicated by the dashed line.

The calculations in the Appendix to Section 6 show that for a satellite in low earth orbit, a significant change in altitude requires a relatively small ΔV. For example, maneuvering from a circular orbit at 400 km to a circular orbit at 1,000 km requires a total ΔV of only 0.32 km/s. On the other hand, if the satellite were transferred from a 400 km orbit to a geosynchronous orbit at 36,000 km altitude, this maneuver would require a total ΔV of 3.9 km/s. Not surprisingly, the ΔV required to change from one circular orbit to another is related to the difference in orbital speeds of the two orbits. Since the orbital speed of circular orbits changes relatively slowly with altitude, orbital changes do not require large values of ΔV unless the change in altitude is very large. This is because the orbital speed is related not to the altitude of the satellite (its distance above the Earth), but to the satellite's distance from the center of the Earth. A relatively large fractional change in altitude, say from 500 to 1,000 km (a 100% change), represents only a small fractional change in the distance to the center of the Earth, in this case from 6,870 to 7,370 km, a 7% change; as a result, the orbital speed changes by less than 4%.

Changing the Orbital Period

Since the orbital period of a satellite depends on the altitude and shape of the orbit, maneuvers to change the shape and altitude of the orbit can be used to change the period. Such maneuvers may be useful, for example, to vary the revisit time of a reconnaissance satellite, making it less predictable. The equation for the change in period produced by a change in velocity is given in the Appendix to Section 6. As an example, a satellite in a circular orbit with an altitude of 400 km has an orbital speed of 7.67 km/s and a period of 92.2 minutes. Increasing the orbital speed by 0.1 km/sec would increase the period by about 3.6 minutes, while an increase of 0.3 km/sec would increase the period by 10.8 minutes. As discussed above, these velocity changes would cause the orbit to become elliptical: the resulting apogees would have altitudes of 750 km and 1,460 km, respectively, while the perigee would remain at 400 km.

Changing the Relative Location of Satellites in the Same Orbit

Changing the period of one satellite can change its position relative to other satellites in the same orbit through a multi-step process. Consider, for example, two satellites in the same circular orbit. Since they must have the same speed, the distance between them will stay the same as they move around the orbit. To change the distance between them, simply increasing the speed of one of the satellites will not work, since that would change its orbit.

Instead, one satellite can be moved relative to the other by putting it temporarily into a higher or lower orbit to change its period, and then moving it back into the original orbit after enough time has passed to put the satellites in the desired relative positions. The amount of propellant required for this process depends on how quickly the change must be made: a small ΔV leads to a small change in period, and the satellites require a long time to reach the desired relative position. For example, consider two satellites that are near one another in a circular orbit at an altitude of 400 km. Giving one satellite a ΔV of 0.1 km/s to place it on an elliptical orbit changes its period by 3.6 minutes, requiring about 13 orbits, or 20 hours, to move it halfway around the orbit relative to the second satellite, which remains on the original orbit. Moving the first satellite back onto the original circular orbit requires another ΔV of 0.1 km/s, for a total ΔV of 0.2 km/s. Doubling the amount of ΔV cuts the transition time roughly in half since it changes the period of the satellite by twice as much (7.2 minutes) as in the previous example.

This type of maneuvering can be used to rendezvous one satellite with another. It can also be used to position multiple satellites around an orbit, as discussed below, to increase the ground coverage of a satellite constellation. These satellites can be placed in the same orbit by a single launcher, then shifted around the orbit by this kind of maneuver.

Maneuvers that Change the Orbital Plane

Maneuvers that change the plane of the orbit require changing the direction of the velocity of the satellite. Since the orbital velocity of a satellite is very large (it varies from roughly 3 to 8 km/sec for typical orbits—see Table 4.1), changing its direction by a significant amount requires adding a large velocity component perpendicular to the orbital velocity. Such large changes in velocity require large amounts of propellant.

Figure 6.3 shows an example for a satellite in a 500 km-altitude orbit, with an orbital velocity of 7.6 km/sec. The figure illustrates that a ΔV of 2 km/s rotates the orbital velocity by only 15 degrees.

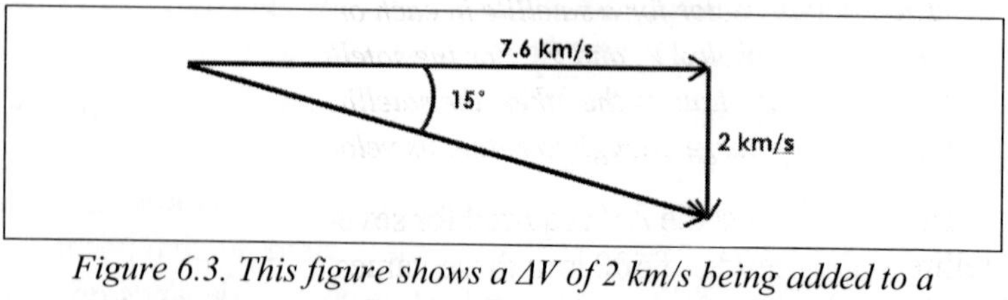

Figure 6.3. This figure shows a ΔV of 2 km/s being added to a speed of 7.6 km/s, which is the orbital speed of a satellite in a 500 km-altitude orbit. Even this large ΔV will lead to only a relatively small change, rotating the orbital plane by an angle of only 15°.

Figure 6.3 shows that the larger the satellite's velocity, the larger the value of ΔV required to rotate the velocity by a given angle. As a result, changing the plane of a satellite in a low altitude circular orbit will require more ΔV than the same change at higher altitudes, because satellites travel at a slower velocity at higher altitudes.

It is convenient to look at two different types of plane-changing maneuvers: those that change the inclination of the plane, and those that rotate the plane at constant inclination. Recall that the orbital plane is partly described by its inclination angle θ, which is measured with respect to the Earth's equatorial plane (see Fig 4.3).

Maneuvers to Change Inclination

The simplest type of plane change to conceptualize is one that changes the inclination of the orbital plane by an angle Δθ. Such a maneuver requires rotating the velocity vector of the satellite by the same angle Δθ (see Figure 6.4).[2]

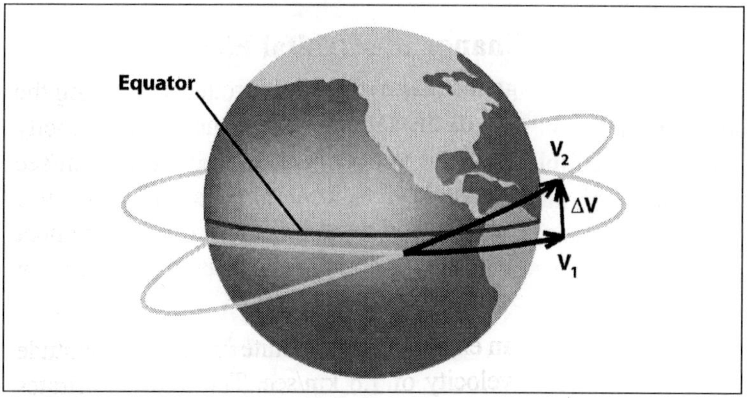

Figure 6.4. This figure shows two orbits with different inclinations. The velocity vector for a satellite in each orbit is denoted by the arrows labeled V_1 and V_2. For the satellite to change its orbit from one plane to the other, the satellite's thrusters must produce a ΔV large enough to rotate its velocity from V_1 to V_2.

Table 6.2 shows the ΔV required for several values of Δθ for a satellite at an altitude of 500 km; these values are calculated using Equation 6.13 in the Appendix to Section 6.

2. This can be thought of as rotating the plane about the line formed by the intersection of the orbital plane and the equatorial plane.

Table 6.2. This table shows values of ΔV required to change the inclination angle θ by an amount $\Delta\theta$ for a satellite at an altitude of 500 km.

Δθ (degrees)	ΔV (km/s)
15	2.0
30	3.9
45	5.8
90	11

Since the orbital speed decreases with altitude, the ΔV required for a given change of Δθ also decreases with orbital altitude, but the decrease is relatively slow. For example, for orbits at 1,000-km altitude, the required ΔV is only 3% lower than for orbits at 500 km (Table 6.2). On the other hand, the required ΔV at geosynchronous altitude (36,000 km) is about 40% of the value of ΔV at 500 km.

For this reason, rotations are made at high altitudes when possible. For, example, consider a satellite that is intended for an equatorial orbit (zero inclination) at geosynchronous altitude, but is launched into a plane with a nonzero inclination due to the location of the launch site. The satellite is placed in an orbit at geosynchronous altitude with nonzero inclination before the orbit is rotated to have zero inclination.

Since the ΔV required for a given Δθ decreases when the satellite's speed decreases, large rotations of the orbital plane can be made somewhat more economically using a three-step process. First the satellite is given a ΔV to increase its altitude at apogee. Since the satellite's speed is slower at apogee, it is rotated at that altitude, then given a final ΔV to reduce the altitude at apogee to its original value. As discussed above, maneuvers that change the altitude require relatively small values of ΔV; consequently, this three-step procedure can, in some cases, require a lower overall ΔV than simply rotating the plane of the original orbit. However, this procedure can take much longer than a simple inclination change because it takes time for the satellite to move into a higher orbit and then return.

As an example, consider a satellite in a circular orbit at an altitude of 500 km. For inclination changes of $\Delta\theta$ less than about 40°, changing altitudes before rotating requires more ΔV than rotating at the original altitude. However, for rotations through larger angles, changing altitude first requires less energy. For example, if $\Delta\theta$ is 90°, performing the rotation at an altitude of 10,000 km reduces the total required ΔV to 8.2 km/s, or 76% of the 10.8 km/s required for such a rotation at the original 500 km altitude. In this case, the total transit time to and from the higher altitude is about 3.5 hours. Rotating instead at an altitude of 100,000 km reduces the required ΔV by nearly 40% to 6.6 km/s and increases the transit time to 37 hours. Going to even higher altitudes reduces the required ΔV only marginally while further increasing transit time.

Rotating the Orbital Plane at Constant Inclination

Another maneuver that can require a large velocity change is rotating the orbital plane around the Earth's axis while keeping the inclination fixed.[3] Such a maneuver might be used if multiple satellites were put into orbit by a single launch vehicle and then moved into different orbital planes—all with the same inclination—to increase the ground coverage of the constellation. A set of three satellites, for example, might be maneuvered to place each in a plane rotated 120° with respect to the others. The energy requirements of such maneuvers are an important consideration when planning to orbit a constellation of satellites.

The ΔV required for this maneuver depends on the angle $\Delta\Omega$ through which the orbital plane is rotated around the Earth's axis, as well as the inclination angle θ of the orbit and the altitude (and therefore the speed) of the satellite when the maneuver is carried out.

Table 6.3 shows the ΔV required for a satellite in a circular orbit at an altitude of 500 km for several rotation angles $\Delta\Omega$ and two inclination angles θ. For practical applications the rotation angle can be large, resulting in very large values of ΔV. As above, the required ΔV decreases slowly with the altitude of the orbit;

3. This can be thought of as rotating the line formed by the intersection of the orbital plane and the equatorial plane about the Earth's axis, while keeping the inclination fixed.

values for a 1,000 km-altitude orbit are about 3% lower than those for a 500 km orbit.

Table 6.3. This table shows the ΔV required for rotations of ΔΩ degrees of an orbital plane around the Earth's axis, for inclinations θ of 45 and 90 degrees. These values assume the satellite is in a circular orbit at an altitude of 500 km.

	θ = 45°	θ = 90°
ΔΩ (degrees)	ΔV (km/s)	ΔV (km/s)
45	4.1	5.8
90	7.6	10.8
120	9.3	13.2

De-Orbiting Maneuvers

For some missions, an object in orbit will use its thrusters to accelerate out of orbit and back toward the Earth. The Space Shuttle must do this to return to Earth; similarly, an orbiting weapon intended to strike the Earth would need to carry propellant to kick it out of orbit. The ΔV required for this maneuver will depend on how fast the return to Earth must be. The dynamics of the deorbiting are complicated because once the satellite moves to low enough altitudes, the increasing density of the atmosphere affects its trajectory.[4]

Figure 6.5 illustrates the de-orbiting process for three values of ΔV. This example assumes a relatively high circular orbit—3,000 km—to show the deorbiting trajectories more clearly. At this altitude, the satellite has an orbital velocity of 6.5 km/s. In this illustration, a thrust is applied instantaneously at point P in a direction opposite to the satellite's velocity, so that it reduces the velocity by ΔV. This reduction in speed causes the satellite to follow an elliptical orbit with a perigee below its original altitude. If the perigee is low enough, the orbit will intersect the Earth.

Making the satellite fall vertically to Earth under the influence of gravity requires reducing its orbital speed to zero—a ΔV of 6.5 km/s. In this case, it would take the satellite 19 minutes to fall

4. These effects include drag forces, which slow the object, and lift forces, which are sideways forces and pull the object off its trajectory. At high speeds, both effects can be important.

to Earth and it would strike the Earth at point O in Figure 6.5, directly below the point at which the velocity change occurred (point P).[5]

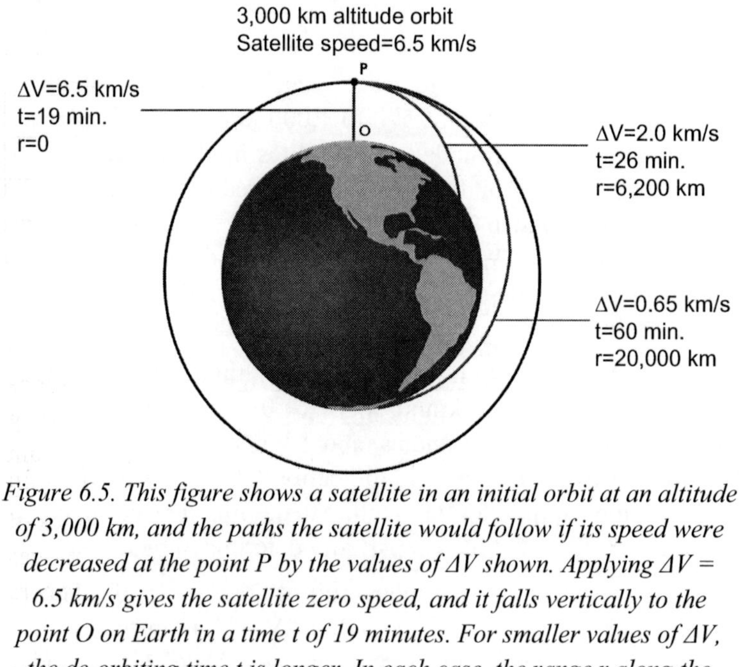

Figure 6.5. This figure shows a satellite in an initial orbit at an altitude of 3,000 km, and the paths the satellite would follow if its speed were decreased at the point P by the values of ΔV shown. Applying ΔV = 6.5 km/s gives the satellite zero speed, and it falls vertically to the point O on Earth in a time t of 19 minutes. For smaller values of ΔV, the de-orbiting time t is longer. In each case, the range r along the Earth's surface is given for the impact point relative to the point O.

Figure 6.5 also shows the reentry trajectory if the satellite's orbital speed were reduced by 2 km/s. In this case, it would take 26 minutes for the satellite to fall to Earth, and it would hit the Earth at a point 6,200 km along the Earth's surface from point O. If the orbital speed were reduced by only 0.65 km/s, so that the satellite takes 60 minutes to de-orbit, it would hit the Earth halfway around the world from point O—at a ground range of roughly 20,000 km.

If ΔV were much less than 0.65 km/s, the satellite would

5. Of course, due to the rotation of the Earth, the point on the Earth that was under the satellite when the ΔV was applied would in general move during the time it took the satellite to reach the Earth; the motion would range from zero at the poles to 500 km at the equator.

not hit the Earth, but would pass by the Earth at low altitude and follow an elliptical orbit to return to point P. However, the drag of passing so low through the atmosphere on its near encounter with the Earth would reduce the satellite's speed, so that it would reach an altitude somewhat less than 3,000 km when it returns to P and would slowly spiral downward on subsequent orbits until it hit the Earth.[6]

A case more relevant to space security issues is a satellite in an orbit with an altitude of 500 to 1,000 km, since this is where missile defense or groundattack satellites might be stationed. In calculating the de-orbit time and ΔV required in this case, assume that the thrust given to the satellite is oriented vertically downward toward the Earth. Applying thrust in this direction results in somewhat shorter de-orbit times than simply reducing the orbital speed as done for the cases illustrated in Figure 6.5.

For a satellite in a circular orbit at an altitude of 500 km (with an orbital speed of 7.6 km/s), a ΔV of 0.7 km/s results in a de-orbit time of about 15 minutes, and 1 km/s in a de-orbit time of 10 minutes (see the Appendix to Section 6 for calculations). (The precise time required for the satellite to de-orbit depends in part on its drag coefficient, which is partially determined by its shape.)

For a satellite in a circular orbit at an altitude of 1,000 km (with an orbital speed of 7.4 km/s), a ΔV of 1.4 km/sec results in a de-orbit time of roughly 15 minutes, and a ΔV of 2 km/sec gives a time of 9 to 10 minutes.

Higher values of ΔV can lead to shorter de-orbit times. Though the satellite would need to carry a large amount of propellant, high ΔVs have been discussed for kinetic energy weapons intended to attack ground targets, which must hit their targets at high speeds. A ΔV of 4 km/s gives de-orbit times of 2 to 3 minutes from an altitude of 500 km, 4 to 5 minutes from 1,000 km, and 14 to 15 minutes from 3,000 km. A ΔV of 6 km/s results in de-orbit times of 1.5 to 2 minutes from an altitude of 500 km, 3 to 3.5 minutes from 1,000 km, and 8.5 to 9.5 minutes from 3,000 km. Section 7 discusses the amount of propellant required for producing these values of ΔV.

6. The object may also be able to use lift forces to assist in de-orbiting, so that the trajectory need not simply be determined by the object's speed.

Reentry Heating

An important issue in de-orbiting is that as the atmosphere slows the satellite large amounts of heat build up in the layers of air around the satellite. (This occurs as the kinetic energy of the satellite is converted to thermal energy of the air, largely through compression of the air in front of the satellite.)

If the object is not to burn up during re-entry, it must carry a heat shield to withstand this intense heat. The heating rate increases rapidly with the speed of the object moving through it and with the density of the atmosphere. If de-orbiting occurs too fast, the satellite will be moving at high speeds low in the atmosphere where the atmospheric density is high, and this can lead to extreme heating.

Atmospheric heating is important when considering the possibility of delivering kinetic energy weapons either from space or by ballistic missile. The motivation for such weapons is that their destructive power would come from the kinetic energy resulting from their high speed rather than from an explosive charge. To be effective, such weapons must hit the ground with very high speed. For example, a mass must be moving at about 3 km/s for its kinetic energy to be equal to the energy released in the explosion of an equal mass of high explosive.[7] The heat load on an object traveling faster than 3 km/s at atmospheric densities near the ground is very large. For comparison, a modern U.S. nuclear reentry vehicle, which is designed to pass through the atmosphere quickly to improve its accuracy, has a speed of about 2.5 km/s when it reaches the ground: designing the warhead to travel faster is limited by its ability to withstand the heating. A penetrator made of a tungsten rod would be more heat tolerant than a nuclear warhead, but the intense heating at the tip of the rod could reduce its structural strength. Since an object traveling at 5 km/s would have a heating rate eight times as high as an object traveling at 2.5 km/s, a kinetic energy weapon traveling at 5 km/s would have to

---

7. The energy released by TNT is roughly 1,000 calories per gram, which equals 4.2 × 106 J/kg. The kinetic energy of a one kilogram mass moving at 3 km/s is V2/2 = 4.5 × 106 J/kg.

withstand eight times the heating rate that a modern U.S. nuclear warhead is designed to tolerate.

Not only do atmospheric forces cause drag, which leads to heating, they can also produce strong lateral forces—called lift forces—that change the object's trajectory. The reentering body can be designed to use the significant lift forces resulting from its high speed in the atmosphere to maneuver in directions perpendicular to its trajectory. Documents describing the goals for ground-attack weapons state that these weapons should be able to travel thousands of kilometers in these directions using only lift forces.

Station Keeping

A number of forces act on a satellite to change its orbit over time. These include the slight asymmetries in the Earth's gravitational field due to the fact that the Earth is not completely spherically symmetric; the gravitational pull of the Sun and Moon; solar radiation pressure; and, for satellites in low earth orbit, atmospheric drag.

As a result, the satellite must periodically maneuver to maintain its prescribed orbit. Thus, it must carry sufficient propellant for this task. While satellite lifetimes used to be limited by the lifetime of the electronics in the satellite, the quality of electronics has improved to the point that lifetimes are now typically limited by the amount of propellant carried for stationkeeping.[8]

How much propellant is needed for stationkeeping depends on several factors. First, satellites that travel for all or part of their orbit at low altitudes (up to several hundred kilometers) must compensate for more atmospheric drag than those at high altitudes. This is especially necessary during high solar activity when the outer parts of the Earth's atmosphere expand, resulting in increased drag at a given altitude. Second, the orbits of some satellites must be strictly maintained, either to fulfill their missions or because their orbital locations are governed by international agreements. For example, the locations of satellites in geosynchronous orbits are tightly controlled by international rules to prevent satellites from interfering with one another. Third, the propellant required depends on the type of thrusters used for stationkeeping, and their

8. Bruno Pattan, Satellite Systems: Principles and Technologies (New York: Van Nostrand Reinhold, 1993), 36.

efficiency. Until recently, conventional chemical thrusters were used for stationkeeping, but other options that reduce propellant requirements are now available. For example, ion thrusters, which provide lower thrust over longer times, are discussed in Section 7.

To get a rough sense of how much maneuvering is required for stationkeeping in geostationary orbits, consider the Intelsat communication satellites. Each year, these use an amount of propellant equal to roughly 2 to 2.5% of their total initial mass (when placed in orbit) for stationkeeping. Thus, for a ten-year satellite lifespan, a propellant mass of 20% to 25% of the satellite's initial mass is required for stationkeeping, which corresponds to a total ΔV over ten years of roughly 0.5–1.0 km/s (see the Appendix to Section 6).

Appendix - 6

Technical Details of Maneuvering

A satellite in a circular orbit at altitude h will have a velocity $V_h^c = \sqrt{GM_e/(h+R_e)}$, where G is the gravitational constant, M_e is the mass of the Earth ($GM_e = 3.99 \times 10^{14}$ m^3/s^2), and Re is the average radius of the Earth (6,370 km) (see the Appendix to Section 4). If the speed of the satellite is suddenly increased by ΔV at some point on the orbit (without changing the direction of the velocity), the orbit becomes an ellipse. The perigee of the new orbit remains at altitude h. The altitude at apogee depends on the value of ΔV. For small ΔV (i.e., ΔV/V << 1), the change in altitude h at apogee is given approximately by[1]

$$\Delta h \approx 4(h+R_e)\frac{\Delta V}{V} \tag{6.1}$$

This equation can be rewritten using $r \equiv (h + R_e)$ as

$$\frac{\Delta r}{r} \approx 4\frac{\Delta Y}{V} \tag{6.2}$$

which shows that the fractional change in r at apogee is just four times the fractional change in the velocity at perigee. Similarly, if the speed of a satellite on a circular orbit is reduced at some point on the orbit, that point becomes the apogee of an elliptical orbit, and the altitude at perigee is less than the altitude of the original orbit by an amount given by Equations 6.1 and 6.2. Equation 6.1 shows why maneuvers that change altitudes take relatively little ΔV: since the change in velocity is multiplied by

1. R. Bate, D. Mueller, and J. White, Fundamentals of Astrodynamics (New York: Dover, 1971), 163.

the radius of the Earth, even a relatively small change in velocity will lead to a significant change in h. This is especially true for satellites maneuvering between low earth orbits, since the altitude band of interest—about 1,000 km—is small compared to Re. For a satellite orbiting at an altitude of 400 km, a ΔV of 0.1 km/s would lead to a change in altitude at apogee of 350 km. If the original orbit is not circular, but elliptical with eccentricity e, the approximate equations for the change in the altitude of the orbit at apogee (Δha) and perigee (Δhp) that result from a velocity change applied at perigee (ΔVp) and at apogee (ΔVa) are, respectively[2]

$$\Delta h_a \approx \frac{4a^2}{GM_e} V_p \Delta V_p = 4\,\frac{h_a + R_e}{1-e}\,\frac{\Delta V_p}{V_p} \quad \text{or} \quad \frac{\Delta r_a}{r_a} \approx \frac{4}{1-e}\,\frac{\Delta V_p}{V_p} \qquad (6.3)$$

and

$$\Delta h_p \approx \frac{4a^2}{GM_e} V_a \Delta V_a = 4\,\frac{h_p + R_e}{1+e}\,\frac{\Delta V_a}{V_a} \quad \text{or} \quad \frac{\Delta r_p}{r_p} \approx \frac{4}{1+e}\,\frac{\Delta V_a}{V_a} \qquad (6.3)$$

Note that these equations are only valid for ΔV/V << 1. For larger values of ΔV, the exact equations given below are required.

Maneuvering between Circular Orbits

Here we calculate the minimum ΔV required to increase the altitude of a circular orbit from h_1 to h_2, through a two-step process using a Hohmann transfer orbit. The transfer orbit is an ellipse with its perigee at h_1 and apogee at h_2 and eccentricity $e = (r_2 - r_1)/(r_2 + r_1)$, where $r_1 \equiv h_1 + R_e$

The first step is to move the satellite from the initial circular orbit onto the transfer orbit by increasing the speed of the satellite from its initial circular velocity $V_1^c = \sqrt{GM_e / r_1}$ to $V_p = V_1^c \sqrt{1+e}$, where e is the eccentricity of the transfer ellipse. This gives

$$\Delta V_p \equiv V_p - V_1^e = V_1^e\left(\sqrt{1+e} - 1\right) \qquad (6.5)$$

The speed of the satellite at apogee of the transfer orbit is $V_a = V_2^c\left(\sqrt{1-e}\right)$, where $V_2^c = \left(\sqrt{GM_e / r_2}\right)$ is the velocity of a circular orbit at altitude h_2. The second step is to make the satellite's orbit circular by increasing the speed at apogee to V c_2. This gives

2. Bate et al., 163.

$$\Delta V_a \equiv V_2^c - V_a = V_2^c\left(1-\sqrt{1-e}\right) \tag{6.6}$$

The total ΔV required for this orbit change is just the sum of these two:

$$\Delta V_{tot} = \Delta V_p + V_a \tag{6.7}$$

For relatively small altitude changes, so that e << 1, this becomes

$$\Delta V_{tot} \approx e\frac{\left(V_1^c + V_2^c\right)}{2} \tag{6.8}$$

Equation 6.8 shows that maneuvering from a circular orbit at 400 km to a circular orbit at 1,000 km requires ΔVtot = 0.32 km/s (in this case, e = 0.041 for the transfer orbit). Moving the satellite from a 400 km orbit to a geosynchronous orbit at 36,000 km altitude requires using a transfer orbit with e = 0.71, so Equation 6.8 cannot be used; Equation 6.7 gives ΔVtot = 3.9 km/s. Two other useful approximate expressions are those for the speed of a satellite at perigee and apogee after a small change of a circular orbit with radius r to an elliptical orbit with semi-major axis of length r + Δr:

$$V_P \approx V^c\left(1+\frac{1}{4}\frac{\Delta r}{r}\right) \text{ and } V_a \approx V^c\left(1-\frac{3}{4}\frac{\Delta r}{r}\right) \tag{6.9}$$

where Vc is the speed of the satellite on the original circular orbit.[3]

Changing the Period of a Satellite

From Equation 4.5 for the period of an elliptical orbit with major axis a

$$\frac{\partial P}{\partial a} = \frac{3}{2}\frac{P}{a} \tag{6.10}$$

and from Equation 4.4 for the speed of a satellite on an elliptical orbit

$$\frac{\partial a}{\partial V} = \frac{2a^2 V}{GM_e} \tag{6.11}$$

Combining these expressions, the change in period ΔP, for small eccentricities, is given approximately by

3. Oliver Montenbruck and Eberhard Gill, Satellite Orbits (New York: Springer-Verlag, 2000), 47.

$$\frac{\Delta P}{P} \approx 3\frac{\Delta V}{V} \tag{6.12}$$

for ΔV/V << 1.

Changing the Inclination of the Orbit

Changing the inclination angle of an orbit by an angle Δθ requires rotating the velocity vector of the satellite by Δθ. Vector addition shows that the required ΔV is

$$\Delta V = 2V \sin\frac{\Delta\theta}{2} \tag{6.13}$$

where V is the speed of the satellite when the maneuver occurs. For circular orbits, the required ΔV decreases with orbital altitude, since the orbital speed decreases with altitude; in this case, ΔV is proportional to $1/\sqrt{h+R_e} \equiv 1/\sqrt{r}$.

Rotating the Orbital Plane at Constant Inclination

For circular orbits, the ΔV required to rotate an orbital plane with an inclination angle θ by an angle ΔΩ around the Earth's axis is

$$\Delta V = 2V \sin\theta \sin\frac{\Delta\Omega}{2} \tag{6.14}$$

where V is the speed of the satellite when the maneuver occurs.[4] As with the previous maneuver, the required ΔV decreases with the altitude of the orbit, since V does. This process is also known as changing the right ascension of the ascending node.

GENERAL ROTATIONS

For circular orbits, the ΔV required for a maneuver that both changes the inclination by Δθ and rotates the orbital plane by an angle ΔΩ around the Earth's axis is given by[5]

$$\Delta V = 2V\sqrt{\sin^2\frac{\Delta\theta}{2} + \sin\theta_1 \sin\theta_2 \sin^2\frac{V\Omega}{2}} \tag{6.15}$$

where V is the speed of the satellite when the maneuver occurs, θ1 and θ2 are the initial and final values of the inclination, and $\Delta\theta \equiv \theta_1 - \theta_2$. Notice that this equation reduces to Equations

4. Vallado, 333. Using trigonometric identities, the equation given in this book can be put in the simpler form given here.
5. Vallado, 335.

6.13 and 6.14 for $\Delta\Omega = 0$ and $\Delta\theta = 0$, respectively. As with the previous maneuvers, the required ΔV decreases with the altitude of the orbit.

De-Orbiting

De-orbiting times and trajectories were calculated using a computer program that integrates the equations of motion for an object, assuming a round Earth with an atmosphere. We assumed the satellite was initially in a circular orbit at altitude h. A velocity change vector of magnitude ΔV was added to the orbital velocity vector, with the change pointing either opposite to the velocity vector or in a vertical direction pointing toward the Earth. We repeated the calculation using a range of drag coefficients for the object, but assumed no lift forces. The drag coefficient enters the calculations through the combination mg/(CdA) called the ballistic coefficient, where m is the mass of the object, g is the acceleration of gravity at the altitude of the object, Cd is the drag coefficient, and A is the cross-sectional area of the object perpendicular to its motion.

In particular, we varied the ballistic coefficient by a factor of 10 from a value comparable to a modern strategic warhead (150,000 Newtons/m2 (N/m2), or 3,000 lb/ft2), to a value for an object with much higher drag (15,000 N/m2, or 300 lb/ft2). As an illustration, consider the case in which the velocity change vector is oriented in the vertical direction. Results are given in Table 6.4.

Table 6.4. This table lists the de-orbiting time for a satellite in a circular orbit at the given altitude when a velocity change ΔV is applied in the vertical direction. The results are given for two different values of the ballistic coefficient, which is inversely proportional to the drag coefficient of the object; the larger value is comparable to that of a modern strategic ballistic missile warhead.

Altitude (km)	ΔV (km/s)	De-orbiting Time (min)	
		Ballistic coefficient 150,000 N/m2 (3,000 lb/ft2)	Ballistic coefficient 15,000 N/m2 (300 lb/ft2)
500	0.7	14.6	15.2

	1	9.4	10.3
	2 4.	4	5.5
	4	2.1	2.9
1,000	1.4	14.4	15.3
	2	9.1	10.2
	4	4.3	5.1
	6	2.8	3.4
3,000	4	14.0	15.1
	6	8.7	9.4

The heating rate for an object moving through the atmosphere is roughly proportional to ρV3, where ρ is the atmospheric density.[6] This expression shows that the heating rate increases rapidly with velocity and with decreasing altitude, since the atmospheric density increases roughly exponentially with decreasing altitude.

Station Keeping

Data from the Intelsat communication satellites suggest the scale of the ΔV required for stationkeeping in geosynchronous orbit using conventional thrusters.[7] The Intelsat V satellite has a mass of 1,005 kg when placed in orbit, of which 175 kg is propellant (with a specific impulse of 290 to 300 s), intended for a lifetime of 7 years. The propellant mass is 17.4% of this initial mass; assuming all the propellant is used for stationkeeping, this corresponds to 2.5% of the initial mass used per year. The Intelsat VII has a mass of 2,100 kg when placed in orbit, of which 650 kg is propellant (with a specific impulse of 235 s) and a planned lifetime of 17 years. The propellant is 31% of the initial mass, and 1.8% is used each year. This indicates that these satellites use roughly 2 to 2.5% of their initial mass per year for stationkeeping. Over a 10-year lifespan, this would require that 20 to 25% of the initial mass be propellant reserved for stationkeeping. Using the rocket equation (see Section 7), these masses can be shown to correspond to a total ΔV over 10 years of roughly 0.5–1.0 km/s.

---

6. For a more detailed discussion of heating at hypersonic speeds, see John Anderson, Hypersonic and High Temperature Gas Dynamics (New York: McGraw-Hill, 1989), 291.
7. Robert A. Nelson, "Rocket Science: Technology Trends in Propulsion," Via Satellite, June 1999, http://www.aticourses.com/rocket_tutorial.htm, accessed January 20, 2005.

Section - 7

Implications of Maneuvering for Satellite Mass

The previous section showed that the laws of orbital mechanics determine the amount of velocity change ΔV required for a satellite to carry out various types of maneuvers (see Table 6.1). This section shows that the mass of propellant a satellite needs to change its speed by ΔV increases rapidly with ΔV. Placing a heavy object in orbit is technologically more difficult than placing a light object. Moreover, the cost of putting a satellite into orbit increases roughly in proportion to its overall mass. These launch factors place practical limits on the amount of propellant a satellite can carry and thus on the amount of maneuvering it can carry out.

The relationship between maneuvering and propellant mass has important implications for space missions such as the proposed military space plane. The space plane is envisioned as a vehicle that, after being launched into orbit, maneuvers to accomplish a variety of tasks.

These might include placing satellites or ground-attack weapons in orbit or rendezvousing with satellites to inspect, repair, or refuel them. However, limits on the mass of propellant that can be launched with such a vehicle places strict limits on how much maneuvering the vehicle could do.

Similarly, the amount of maneuvering a reconnaissance satellite could do for either offensive or defensive purposes is limited by the amount of propellant it carries. Section 9 discusses some of these consequences.

Section 6 showed that the ΔV required for a particular maneuver is determined by physics. However, the propellant mass required to provide that ΔV depends on how efficiently the thruster can use propellant to bring about a velocity change, which depends on the thruster technology.[1] Most of the calculations in this paper assume conventional thrusters using chemical propellant, for reasons explained below.[2] Other thruster technologies are also discussed, along with their applications and limitations.

Satellite Mass

The relationship between ΔV, the mass of propellant Mp needed to impart ΔV to this satellite, and the satellite mass Ms (which does not include Mp)[3] is given by the so-called rocket equation, which the Appendix to Section 7 discusses in detail.

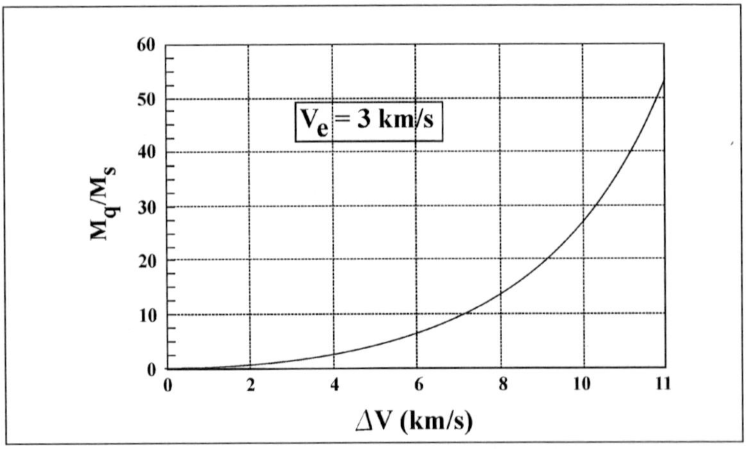

Figure 7.1. This figure shows the ratio of propellant mass to satellite mass Mp/Ms required to produce a given velocity change (ΔV) assuming conventional propulsion (Ve = 3 km/s). Ms is the mass of the satellite excluding the propellant mass Mp.

Table 7.1 lists several values of ΔV and the corresponding values of Mp/Ms, assuming conventional propulsion technology. (The equations used to calculate these values are given in the

1. The thruster efficiency is sometimes expressed in terms of its specific impulse. It can be expressed equivalently as the exhaust velocity Ve of the particles ejected from the thruster, which is the term used in this paper.

2. We use a value of Ve = 3 km/s for conventional thrusters.

3. The mass Ms of the satellite may include propellant for purposes such as stationkeeping and maneuvers other than the one being considered here.

Appendix to Section 7.) For example, to carry out a maneuver requiring a ΔV of 2 km/s, the propellant mass Mp required for this maneuver is 0.9 times that of the satellite itself, that is, the propellant nearly doubles the total mass that must be placed in orbit. In other words, a satellite with a mass of one ton (excluding the propellant for this maneuver) would need to carry 0.9 tons of propellant to provide the ΔV for this maneuver. This mass penalty increases rapidly as ΔV increases, as Figure 7.1 and Table 7.1 show. To carry out a maneuver requiring a ΔV of 5 km/s (or several maneuvers that added up to 5 km/s), a one-ton satellite would need to carry 4.3 tons of propellant to conduct this maneuver. For a maneuver (or set of maneuvers) requiring a ΔV of 10 km/s, that same satellite would need to carry 27 tons of propellant.

Table 7.1. Selected values of the ratio of propellant mass to satellite mass Mp/Ms required to produce a given velocity change (ΔV) assuming conventional propulsion (Ve = 3 km/s).

ΔV (km/s)	Mp/Ms
1	0.4
2	0.9
3	1.7
4	2.8
5	4.3
8	13.4
10	27.0
12	53.6

Note that the propellant mass needed to deliver a given ΔV and carry out a given maneuver depends on the total satellite mass at the time the maneuver takes place. Therefore, if a satellite is carrying a large amount of propellant for multiple maneuvers, the initial maneuvers require proportionately more propellant since the mass that must be accelerated is that of the satellite plus the remaining propellant.

Table 7.2 gives the ratio of propellant mass to satellite mass required by the maneuvers considered in Section 6 (assuming conventional thruster technology on the satellite). These numbers were calculated using the rocket equation (see the Appendix to Section 7).

Table 7.2. This table gives the ratio of propellant mass to satellite mass M_p/M_s required for the space activities listed in Table 6.1, assuming conventional propulsion (V_e = 3 km/s). M_s is the mass of the satellite excluding the propellant mass M_p.

Type of Satellite Maneuver	Required ΔV (km/s)	M_p/M_s
Changing orbital altitude within LEO (from 400 to 1,000 km)	0.3	0.1
Stationkeeping in GEO over 10 years	0.5–1	0.2–0.4
De-orbiting from LEO to Earth	0.5–2	0.2–1
Changing inclination of orbital plane in GEO		
by Δθ = 30°	2	1
by Δθ = 90°	4	3
Changing orbital altitude from LEO to GEO (from 400 to 36,000 km)	4	3
Changing inclination of orbital plane in LEO		
by Δθ = 30°	4	3
by Δθ = 90°	11	38

The Potential Impact of New Technologies

How much propellant the satellite needs to produce the necessary ΔV depends on the thruster technology. The basic physics of rocket thrusters of all types is the same: a power source accelerates the propellant material to high speed and ejects it in a specific direction as exhaust, which propels the satellite in the direction opposite to that of the exhaust. The amount of thrust produced by this process depends on the speed of the particles in the exhaust (the exhaust velocity) and the amount of mass the thruster can eject every second (the mass flow rate).

The mass ratios given in the tables above are for conventional thruster technologies fueled by chemical propellants. These are by far the most prevalent and will remain so for many applications.

New technologies that use propellant more efficiently are being developed, including electric arcjet thrusters and electric ion thrusters. However, these systems produce much less thrust than conventional thrusters. While these thrusters require less propellant mass to produce a given ΔV, the thruster must operate for a much longer time to produce that ΔV. Ion thrusters, for example, have exhaust velocities 10 to 20 times higher than chemical thrusters, but currently their mass flow rates are many thousands of times smaller. As a result, these engines produce thrust levels thousands of times less than conventional thrusters, which would require them to operate thousands of times longer than a conventional engine to bring about the same ΔV (see below).

Such thrusters can be practical for an application such as stationkeeping, which does not need to occur rapidly. But low-thrust engines are not appropriate for missions that require a rapid response, such as ballistic missile defense and other military missions.

For such applications, chemical thrusters remain the only practical choice for the foreseeable future. For the longer term, NASA is considering various types of nuclear propulsion for its spacecraft. While such engines may produce higher thrust, they are unlikely to be used near the Earth for the applications of interest in this report.

Below we provide more information on conventional thrusters and several new thruster technologies.[4]

Conventional Thrusters

In a conventional thruster, the power source is a chemical reaction, which heats the propellant to high temperatures. A nozzle directs the hot gases so that they provide thrust efficiently. Conventional chemical thrusters have moderate values of exhaust velocity (up

4. For an overview of electric propulsion technologies, see Alec D. Gallimore, "The Past and Future of Rocket Engine Technologies," http://www.fathom.com/course/21701743/index.html, accessed January 20, 2005.

to 3 or 4 km/s), but can have large mass flow rates that give rise to large thrust forces. For example, conventional thrusters used on satellites produce thrusts of several hundred or even several thousand Newtons (N).[56]

Electric Arcjet

Thrusters This technology provides a way of improving the efficiency above that of conventional thrusters and increasing the exhaust velocity to above 5 km/s. These systems use an arcjet to superheat the propellant before it is burned, which increases the efficiency of the process. However, as with ion thrusters, the thrusts that can currently be produced by these systems are small—less than a Newton. Such thrusters were first used on satellites for stationkeeping in 1993. Compared with a conventional thruster with an exhaust velocity of 3 km/s, a thruster with an exhaust velocity of 5 km/s could reduce the amount of stationkeeping propellant required[7] by 40%.

Ion Thrusters

The main alternative to conventional propulsion now being developed is electric ion thrusters. A number of varieties are under development, but all work on the principle of creating charged ions that are accelerated to high speed by an electric field. This method can produce high exhaust velocities—values 10 to 20 or more times those provided by conventional thrusters have been achieved. However, the mass flow rate is many thousands of times smaller than that produced by conventional thrusters, and their thrust levels are still typically less than a Newton. An ion thruster was first used on a commercial satellite in 1997.[8]

An example of the ion engine is the Xenon-Ion Propulsion System (XIPS) engines used on Boeing 702 communications

5. A Newton is a unit of force, with dimensions of kg-m/s2.

6. For examples of thrusters developed for this use, see the EADS Space Transportation website http://cs.space.eads.net/sp/, accessed January 20, 2005. An example of a large upper stage engine is the Fregat upper stage of the Soyuz vehicle, which has a thrust of 19.4 kN; see Starsem, "Soyuz User's Manual," April 2001, http://www.starsem.com/services/images/ soyuz_users_manual_190401.pdf, accessed January 20, 2005.

7. Nelson.

8. Nelson.

satellites for stationkeeping and changing altitude.[9] It has an exhaust velocity of 35 km/s, which is more than ten times as great as conventional thrusters. However, the mass flow is only 5×10^{-6} kg/s, providing a maximum thrust of 0.165 N.

Each Boeing satellite reportedly operates a set of four of these engines for 30 minutes per day for stationkeeping, using 5 kg of propellant per year, or a total of 50 to 75 kg over the 10- to 15-year lifespan of the satellite.[10] A conventional thruster with a Ve of 3 km/s on a satellite of the same mass would require 10 to 15 times as much propellant mass to provide the same ΔV.

However, ion engines require high power: the XIPS engine uses 4.5 kW of power. This power is supplied by the solar panels of the Boeing 702 satellites, which deliver 10 to 15 kW of power.

The High Power Electric Propulsion (HiPEP) engine currently being developed by NASA is also an ion engine.[11] Because this engine is intended to operate at up to 50 kW and NASA plans to use it in spacecraft that operate far from the Sun, it is being developed as part of the nuclear electric propulsion (NEP) programs under Project Prometheus.

The HiPEP engine has demonstrated high exhaust velocities, but as with other ion engines, produces low thrust. In an initial test in November 2003, the HiPEP engine demonstrated exhaust velocities from 60 to 80 km/s and operated at power levels up to 12 kW.[12] A February 2004 test apparently operated at 34 kW, with

9. Technical information on the XIPS thruster is available on the Boeing website at http://www.boeing.com/defense-space/space/bss/factsheets/xips/xips.html, accessed January 20, 2005 and http://www.boeing.com/ids/edd/ep.html, accessed January 20, 2005.

10. "Boeing 702 Fleet," http://www.boeing.com/defense-space/space/bss/factsheets/ 702/ 702fleet.html, accessed January 20, 2005.

11. NASA Glenn Research Center, "High Power Electric Propulsion Program (HiPEP)," http://www.grc.nasa.gov/WWW/ion/present/hipep.htm, accessed January 20, 2005.

12. NASA, "NASA Successfully Tests Ion Engine," Press Release 03-377, November 20, 2003, http://www.nasa.gov/home/hqnews/2003/nov/HQ_03377_ion_engine.html, accessed January 20, 2005.

an exhaust velocity of 95 km/s. The thrust generated in this test was 0.6 N.[13]

To illustrate the difference in time required for the same maneuver using different thrusters, consider a maneuver that requires a ΔV of 1 km/s. A conventional thruster (with a thrust of 1,000 N and exhaust velocity of 3 km/s) would need to operate for about 4 minutes to execute this maneuver. The current generation XIPS engine would need to operate for 2 weeks, and the advanced HiPEP engine for 4 days (see the Appendix to Section 7 for details).[14]

Nuclear Propulsion

NASA is working, under Project Prometheus, on several projects related to nuclear power in space and is considering several others.[15] A primary goal is to develop electric power sources as alternatives to solar power for spacecraft operating far from the Sun. One project is developing generators containing radioisotopes to produce relatively low levels of electric power (hundreds of watts) for spacecraft. Generators of this type use plutonium-238, and versions have been flown in previous NASA spacecraft. A second project under consideration would develop a uranium-fueled nuclear reactor for use in space that would produce electric power for electric propulsion and other systems on the spacecraft. Current discussions call for a reactor capable of producing 100 kW of electricity.[16]

A longer-term focus is the nuclear thermal propulsion program, which would use the heat from the nuclear reactor to heat a propellant and create high thrust.

---

13. Alan Newhouse, "Project Prometheus: Program Overview," briefing slides, April 29, 2004, http://www.spacecongress.org/2004/Other/Newhouse.pdf, accessed January 20, 2005.

14. This example assumes a satellite mass, without propellant, of 200 kg.

15. Project Prometheus, http://spacescience.nasa.gov/missions/prometheus.htm, accessed March 1, 2004.

16. Ben Iannotta, "Jupiter Moon Probe Goes Nuclear," Aerospace America, March 2004, http://www.aiaa.org/aerospace/Article.cfm?issuetocid=474&ArchiveIssueID=50, accessed January 20, 2005, and Newhouse, "Project Prometheus."

Appendix - 7

The Rocket Equation

One of the most basic and important equations of orbital dynamics is the rocket equation, which relates the mass of propellant required to impart a given ΔV to a satellite of given mass, for a given thruster technology. The equation is a direct consequence of conservation of momentum (see below for derivation). The equation can be written in several useful forms:

$$\Delta V \equiv V_e \ln\left(\frac{M_i}{M_f}\right) \tag{7.1}$$

$$\frac{M_i}{M_f} = e^{\Delta v / v_e} \tag{7.2}$$

where M_i and M_f are the initial and final mass of the satellite before and after the thruster operates, and V_e is the exhaust velocity of the rocket motor providing the thrust, or the average speed at which the mass in the exhaust is ejected from the thruster. In the process of bringing about this maneuver, the thruster burns a mass $M_p = M_i - M_f$ of propellant. The final mass includes the mass of the unfueled satellite and of any additional propellant it carries. From Equation 7.2, the mass of propellant M_p required for a maneuver of ΔV for a satellite can be written in terms of the initial mass M_i, as

$$M_p = M_i (l - e^{-\Delta v / v_e}) \tag{7.3}$$

or in terms of the final mass M_f, as

$$M_p = M_i (e^{\Delta v / v_e} - 1) \tag{7.4}$$

The exhaust velocity is often expressed in terms of specific impulse, Isp:

$$V_e = g_o I_{sp} \tag{7.5}$$

where g_o is the acceleration of gravity at the Earth's surface (9.81 m/s2). The thrust T of a rocket engine is given by

$$T = V_e \frac{dM}{dt} \tag{7.6}$$

where dM/dt is the mass flow rate of propellant.

Equation 7.6 shows that even if Ve is large, the thrust will be small if the mass flow is small. This is the situation discussed for ion thrusters in the text.

where dM/dt is the mass flow rate of propellant.

Equation 7.6 shows that even if Ve is large, the thrust will be small if the mass flow is small. This is the situation discussed for ion thrusters in the text.

Derivation of the Rocket Equation

The rocket equation is a simple consequence of conservation of momentum. If a body of mass M ejects a particle of mass dM at a speed Ve, the body's speed increases in the opposite direction by an amount dV. Momentum conservation requires

$$MdV = -V_e dM \tag{7.7}$$

Repeating this process increases the body's speed and decreases its mass. Solving for dV and integrating gives

$$\int_{V_i}^{V_f} dV = V_f - V_i \equiv \Delta V = -V_e \int_{M_i}^{M_f} \frac{dM}{M} = -V_e (\text{In } M_f - \text{In } M_i) \tag{7.8}$$

$$\Delta V = V_e \ln\left(\frac{M_i}{M_f}\right) \tag{7.9}$$

where Vi and Vf are the initial and final velocities, and Mi and Mf are the initial and final masses.

The Time Required for a Maneuver

Since the rocket equation includes the exhaust velocity of the engine used to create the required ΔV, the propellant mass required to deliver a given ΔV can be reduced by technologies that increase V_e relative to conventional thrusters. However, if the engine produces a small thrust, the time to conduct such a maneuver can be long.

The time required to complete a given maneuver can be found by dividing the mass of propellant needed to provide the ΔV by the mass flow rate of the thruster. Using Equations 7.4 and 7.6 this can be written as

$$\Delta t = \frac{M_p}{dM / dt} = \frac{M_f (e^{\Delta V / V_e} - 1)}{T / V_e} \tag{7.10}$$

Section - 8

Getting Things into Space: Rockets and Launch Requirements

To place an object in orbit, a rocket must be able to do two things: carry the object to the proper altitude and give it the correct speed at that altitude. Even a short-range missile can launch a payload to altitudes of several hundred kilometers (at which point it will fall back to Earth), whereas placing an object into low earth orbit requires a much more powerful rocket.[1] Developing rockets powerful enough to place satellites in orbit is a difficult technical challenge; currently, only a handful of countries have developed this capability.

Using Missiles to Launch Pay Loads to High Altitudes: The " 1 / 2 Rule "

A useful rule of thumb is that a ballistic missile that can launch a given payload to a maximum range R on the Earth can launch that same payload vertically to an altitude of roughly R/2. This relation is exact in the case of a flat Earth and therefore holds for missiles with ranges up to a couple thousand kilometers (the Earth appears essentially flat over those distances, which are small compared to the radius of the Earth). But the rule continues to hold approximately for even intercontinental range missiles (see Appendix B to Section 8).

Changing the payload of the missile changes both its maximum range and its maximum altitude, but these two distances continue

1. For example, the potential energy of a mass lifted to a 300 km altitude is less than 3% of the kinetic energy of the same mass in a circular orbit at that altitude. (See Appendix A to Section 8 for a discussion of the potential and kinetic energies of orbits.)

to be related by the 1/2 Rule. For example, a Scud missile, which has a maximum range of 300 km with a one-ton payload, would be able to launch a one-ton payload vertically to an altitude of 150 km. By reducing the payload, the Scud missile could launch it to higher altitudes. Reducing the payload of a Scud missile by onehalf, to 500 kg, would give it a maximum range of about 440 km or allow it to launch the payload to a maximum altitude of about 220 km. Reducing the payload to 250 kg would increase the maximum range to about 560 km, and the maximum altitude to about 280 km.

Putting Objects into Orbit

Recall from Section 4 that a satellite in low earth orbit has a speed of 7 to 8 km/s. The rocket placing the satellite into orbit must therefore be able to reach that speed.

It is useful to compare this speed to that of ballistic missiles of various ranges.[2] A short-range missile, such as the 300-km range Scud missile that the Soviet Union developed in the 1960s, reaches a top speed of about 1.4 km/s. For a missile to reach a range of 1,000 km, it must be able to reach a speed of 3 km/s. An intercontinental ballistic missile, similar to those the United States and Russia deploy as part of their strategic nuclear forces, is capable of reaching a quarter of the way around the Earth (10,000–12,000 km) and reaches a speed in excess of 7 km/s.

The similarity in speed between an intercontinental missile and a rocket needed for space-launch is the reason that similar technology can be used for both and that countries have generally developed the two capabilities at the same time.

Note, however, that even an intercontinental-range missile cannot place its full payload into orbit. A 10,000-km range missile typically burns out at an altitude of several hundred kilometers. Its

2. A ballistic missile warhead is not powered throughout its flight. Instead the missile rapidly accelerates the warhead to high speed and then releases it, so that for most of its trajectory the warhead is falling through space (this free-falling motion is called ballistic, accounting for the term ballistic missile). The range that a warhead can reach depends on how fast the rocket booster is traveling when it releases the warhead, just as the distance a baseball travels depends on how fast it is moving when it leaves your hand. The rocket booster reaches it maximum speed at "burnout," when the rocket finishes burning its fuel.

speed of roughly 7 km/s is about 10% too low to place a satellite in a circular orbit at that altitude. Reaching the necessary speed would require reducing the payload by roughly a third.[3] This fact is important in comparing space-based and missile-launched weapons, as is done in Section 9.

Placing an object into orbit is thus technically demanding. It is also expensive. A rough rule of thumb is that a modern rocket can deliver into orbit a payload that is only a few percent of the rocket's overall mass. Since the size of the rocket needed to put a satellite into orbit scales with the mass of the satellite, there is a tremendous incentive to keep the mass of satellites as low as possible.

Table 8.1 gives data on several space-launch vehicles, including the lift-off mass of the launcher and the mass that it can place in three different types of orbits: circular low earth orbits with altitudes of a few hundred kilometers; geosynchronous transfer orbits, which are elliptical orbits with perigee typically at a few hundred kilometers and apogee at geosynchronous altitude of approximately 36,000 km; and sun-synchronous orbits, which typically have altitudes below 1,000 km and an inclination near 90°. The table shows that modern launchers have lift-off masses of 200 to 700 metric tons and are able to place 2.5% to 4% of their lift-off mass in low earth orbit. As the table also shows, for a given launcher, the mass that can be placed in a geosynchronous transfer orbit is roughly half the mass that can be placed in low earth orbit; this is comparable to the mass that can be placed in a sun-synchronous orbit.

Table 8.1: This table shows the satellite mass that various space launchers can place in different orbits.[4] M_{LEO}, M_{GTO}, and $M_{SS}O$ are the masses that can be placed in low earth orbit (LEO), geosynchronous transfer orbit (GTO), and sunsynchronous orbit (SSO), respectively. (A metric ton is 1,000 kg.) In the second column, the numbers in parentheses are the altitude h and inclination θ of the orbits. In the

3. Steve Fetter, University of Maryland, personal communication, July 2004. For several different missiles, Fetter calculated the mass that the missile could launch into low earth orbit (200 and 500 km altitude) and compared it with the mass he calculated the same missile could send to 10,000 km range.

4. S.J. Isakowitz, J.P. Hopkins, Jr, and J.B. Hopkins, International Reference Guide to Space Launch Systems, 3rd ed. (Reston, VA: American Institute of Aeronautics and Astronautics, 1999).

third column, these numbers are the perigee of the elliptical orbit, and the inclination θ. In all cases the apogee is at geosynchronous altitude of roughly 36,000 km. The fifth column gives the lift-off mass of the launcher, M_{LO}. The final three columns give the ratio of the satellite mass to the lift-off mass for the three types of orbits.

Launcher	M_{LEO} (metric tons) (h, θ)	M_{GTO} (metric tons) (perigee, θ)	M_{SSO} (metric tons)	M_{LO} (metric tons)	$\frac{M_{LEO}}{M_{LO}}$	$\frac{M_{GTO}}{M_{LO}}$	$\frac{M_{SSO}}{M_{LO}}$
Ariane 4	10.2 (200km, 5.2°)				2.6%		
(AR44L)	8.2 (200km, 90°)	4.8 (185km, 7°)	6.5	470	1.7%	1.0%	1.4%
(Europe)							
Ariane 5	18 (550km, 28.5°)	6.8 (580km, 7°)	12	737	2.4%	0.92%	1.6%
(Europe)							
Atlas IIA	7.3 (185km, 28.5°)				3.9%		
(USA)	6.2 (185km, 90°)	3.1 (167km, 27°)		188	3.3%	1.6%	
Atlas V 550	20 (185km, 28.5°)				3.7%		
(USA)	17 (185km, 90°)	8.2 (167km, 27°)		540	3.1%	1.5%	
Delta III	8.3 (185km, 28.7°)				2.8%		
(USA)	6.8 (200km, 90°)	8.3 (200km, 28.7°)	6.1	302	2.2%	1.3%	2.0%
GSLV	5 (200km, 45°)	2.5 (185km, 18°)	2.2	402	1.2%	0.62%	0.55%
(India)							
H-2	10.6 (200km, 30.4°)	3.9 (250km, 28.5°)	4.2	260	3.9%	1.5%	1.6%
(Japan)							
H-2 A2024	11.7 (300km, 30.4°)	5.0 (250km, 28.5°)	5.3	289	4.1%	1.7%	1.8%
(Japan)							
LM-3B	11.2 (200km, 28.5°)	5.1 (180km, 28.5°)	6.0	426	2.6%	1.2%	1.4%
(China)							
Proton K	19.8 (186km, 51.6°)	4.9 (4200km, 23.4°)	3.6	692	2.9%	0.71%	0.52%
(Russia)							
PSLV	3.7 (200km, 49.5°)	0.8 (185km, 18°)	1.3	294	1.3%	0.3%	0.4%
(India)							

Factors That Affect Launch Capability

The capability of a given launcher to place objects in orbit depends on many factors, as listed below.

- *The mass being lifted into space (the payload).* As a consequence of the rocket equation, the propellant in the missile cannot accelerate a massive payload to as high a velocity, and thus lift it to as high an altitude, as it can a lower mass payload. Moreover, the heavier the payload, the more gravity slows the rocket as it travels to high altitudes. As a result, the more massive the payload, the lower the altitude at which the rocket can place this mass in orbit.

- *The location of the launch site and the direction of the launch.* The rotation of the Earth gives a rocket an eastward velocity even before it is launched.[5] If the rocket is launched to the east, it can use this velocity to increase its speed. Since the speed of the Earth's surface is greatest at the equator (0.456 km/s), launching from a location at low latitudes (near the equator) increases the rocket's speed and therefore increases its launch capability. In addition, for launches into geostationary orbit, launching from near the equator can place the satellite into an orbital plane with the correct inclination; launching from higher latitudes requires the launcher to use propellant to rotate the orbital plane to make it equatorial.

For example, a rocket launched from the French Kourou launch site at 5.23° latitude could carry 20% more mass into a geosynchronous transfer orbit than could the same rocket from the Kazakh Baikonur launch site at 46° latitude.[6] For a launch site at 70° latitude, the rocket could only carry half as much mass as one launched from Kourou.

Similarly, if the rocket is not able to launch eastward, it cannot take full advantage of the speed of the Earth's rotation, and this reduces its launch capability. This can happen, for example, if the satellite is being launched into a polar orbit, in which case the rocket is launched toward the north or south. Or the launch directions may be restricted so that the rocket does not fly over populated areas early in flight. This constraint may impose a fuel-costly orbital maneuver to reach the desired final orbit, and thus reduce the launch capability. Both India and Israel are in this situation. The relatively low mass ratios

5. The additional speed that a launcher can use from the Earth's rotation is given by: $V = 0.456 \cos \theta \cos \alpha$, where θ is the latitude of the launch site and α is the angle between the direction of the launch and due east.

6. The launch from Baikonur would lose 0.14 km/s from the Earth's rotation relative to a launch from Kourou, and rotating the orbital plane from an inclination of 46° to 0° would require $\Delta V = 2.4$ km/s, assuming it was done once the satellite was in geostationary orbit.

shown in the last three columns of Table 8.1 for the Indian GSLV and PSLV rockets reflect both India's somewhat less mature technology and the geographic restrictions on the directions it can launch, which prevent it from taking full advantage of the rotation of the Earth.[7]

Another example is the North Korean attempt to launch a satellite in August 1998, which was launched eastward over Japan. Because the rocket passed over Japan, many saw this act as threatening; however, this trajectory was likely chosen to take maximum advantage of the Earth's rotational speed.

- *Details of the orbit*. The altitude, shape, and inclination of the orbit all affect orbital launch capability. For example, if Vh is the speed a satellite needs to be placed in a circular orbit at altitude h, a higher speed equal to $\sqrt{1+e}\,V_h$ is required for a satellite to be placed into an elliptical orbit (with eccentricity e) at its perigee point at an altitude h.

  As noted above, if the satellite needs to be placed in a highly inclined orbit, it is likely to be launched in a direction that does not allow it to make maximum use of the Earth's rotational speed. This can be illustrated by comparing the masses that can be launched into low earth orbits with different inclinations in Table 8.1. Four of the entries provide data for launches into polar orbits and into orbits with lower inclinations with the same or comparable altitude: Ariane 4, Atlas IIA, Atlas V 550, and Delta III. In each case, the rocket is able to place 20% more mass into the low inclination orbit than the polar orbit.

  Alternately, since a satellite cannot be launched into an orbit with inclination less than the latitude of the launch site (see Section 4), if a satellite is launched from a location in the mid-latitudes but is intended for an orbit with inclination near zero, the satellite must maneuver to change orbital planes, which requires additional fuel mass to be carried into orbit, which reduces the launch capability.

7. Isakowitz et al., 310.

Placing Satellites in Geostationary Orbit

Satellites are typically placed in geostationary orbits in two steps. The first step is to launch the satellite into a parking orbit, which is typically at low altitude (200 to 300 km). The second step is to maneuver the satellite into an elliptical Hohmann transfer orbit, or geosynchronous transfer orbit (GTO), to change the orbit from low earth orbit to geosynchronous orbit (see Figure 6.2). The transfer orbit has its perigee at the parking orbit's altitude and its apogee at geosynchronous altitude. A ΔV of 2.4 km/s is required to place the satellite into GTO from the parking orbit, and another ΔV of 1.5 km/s is required to circularize the orbit at GEO, for a total ΔV of 3.9 km/s for reaching a geosynchronous orbit.

For the satellite to be in a geostationary orbit, it must be in an equatorial (inclination = 0°) geosynchronous orbit. If the inclination of the orbit needs to be changed, this is typically done once the satellite is at synchronous altitude, since it requires less propellant, as discussed in Sections 6 and 7.

To place the satellite in the right location in geostationary orbit, proper timing is required. A direct launch to geostationary orbit would need to be timed for the satellite to reach geostationary altitude at the destination position, or the satellite would need to maneuver to its assigned position once in geostationary orbit. By using a parking orbit, the timing for maneuvering into GTO can be separated from considerations that determine the timing of the launch.

Air Launching

The mass of the launcher needed to place a satellite in orbit roughly scales with the mass of the satellite, as described above. Consequently, launching small payloads requires a small launch vehicle, and this opens up the possibility of using a rocket that can be carried aloft by an aircraft. A small payload may result from future miniaturization of satellite technology. Or the payload may be a relatively simple system, such as a simple interrogation satellite, a small kill vehicle, or a space mine.

Air-launching has a number of practical advantages. Since the launch does not require a dedicated launch facility, this can in principle reduce costs and allow rapid launches.

Since the launcher is mobile, the user can choose the location and latitude of the launch and can reduce restrictions on the direction of launch by, for example, launching over the ocean. This increases the efficiency of getting to orbit and allows a satellite to be launched directly into a desired orbit rather than launching into an orbit determined by the launch site and then maneuvering into the proper orbit.[8]

Since the atmosphere rotates with the Earth, launching eastward from an aircraft allows the launcher to take advantage of the rotational speed of the Earth, just as launching from the ground does. Since the booster is released above the ground and with an initial velocity equal to that of the aircraft, the requirements on the booster are somewhat reduced. For example, some of the configurations discussed below could increase the booster payload by more than 50% relative to that for the same booster launched from the ground.

Pegasus is an existing air-launched booster that is carried aloft by a B-52 for military payloads or by an L-1011 aircraft for civil payloads. The Pegasus XL has a mass of 23 tons. It can place 450 kg into a 200 km orbit at 28° inclination, 330 kg into a 200-km polar orbit (90° inclination), and 190 kg into an 800-km sun-synchronous orbit. The aircraft releases the three-stage Pegasus booster at an altitude of 12 km at a speed of roughly 0.25 km/s.[9] As of August 2003, Pegasus had been used in 35 launches, the first in 1990.[10]

Other air-launch systems are being developed. The Air Force Research Laboratory is developing a microsatellite launch vehicle

8. Small ground-based launchers can have some of these advantages. For example, the SpaceX Falcon 1 launcher is reported to require less launch infrastructure and, therefore, for some missions can be launched from Omelek Island in the Marshall Islands, which lies on the equator. See Craig Covault, "The SpaceX Falcon Will Challenge Orbital Sciences and Boeing," Aviation Week and Space Technology, March 28, 2004, http://www.spacequest.com/Articles/The%20 SpaceX %20Falcon%20Will%20Challenge%20Orbital%20Sciences%20.doc, accessed January 21, 2005.

9. Isakowitz et al., 268–279.

10. Orbital Sciences Corporation, "Pegasus Mission History," http://www.orbital.com/Space Launch/Pegasus/pegasus_history.htm, accessed January 21, 2005.

(MSLV) that would be launched from an F-15E aircraft, although there are currently no plans to build the system. The goal is a three-stage booster that could place a 100-kg satellite into a 225-km orbit. The aircraft is intended to climb at a 60° angle and release the booster at an altitude of 11.6 km at a speed of about 0.5 km/s. Ultimately, the goal is a 5-ton booster that would be able to place up to 200 kg in a 280-km orbit within 48 hours.[11]

The Defense Department is developing a system called RASCAL (Responsive Access Small Cargo Affordable Launch), which is intended to include an aircraft capable of releasing an expendable booster at much higher altitudes. The aircraft is being designed to release a booster of up to 8 tons at an altitude of 60 km and a speed of 0.37 km/s, with a response time of 24 hours. The booster has not been designed, but the goal is to be able to place roughly 200 kg into a 300-km altitude orbit at low inclinations, or roughly 100 kg into an 800-km sun-synchronous orbit. The first two launches are planned for 2006.[12]

Launch Costs

The cost of launching satellites into orbit is generally discussed in terms of launch costs per satellite mass, which assumes that the cost roughly scales with the mass of the satellite. While this is not necessarily true, it is a convenient way to estimate launch costs and to compare costs of different launch vehicles. A typical number given for the cost per kilogram of launching objects into low-earth orbit is roughly \$20,000 per kilogram (\$10,000 per pound).[13] This cost refers to the cost of launching on a large space-launch vehicle such as those shown in Table 8.1.

Not surprisingly, a key goal in developing new launchers is to reduce launch costs. However, there is ongoing debate about

---

11. William Scott, "Fighters as Spacelift," Aviation Week and Space Technology, April 7, 2003, 72.

12. Robert Wall, "Hot Rod to Space," Aviation Week and Space Technology, September 22, 2003, 48.

13. See, for example, American Physical Society (APS), Report of the American Physical Society Study Group on Boost-Phase Intercept Systems for National Missile Defense, July 2003, 127, http://www.aps.org/public_affairs/popa/reports/nmd03.html, accessed January 5, 2005.

what factors drive up launch costs and how best to lower them.[14] A second goal is to reduce the time required to launch a satellite. Rapid response is of particular interest to some in the U.S. military, who talk of reducing the time to launch a satellite from weeks or months to hours or days.[15]

One path of development has been reusable launchers, but these will not be available in the near future, and it is unclear to what extent they may reduce launch costs.

A second path is reducing the cost of building and launching conventional launchers. The company SpaceX states that by 2006 its Falcon V launcher, which is similar in size to an Atlas IIA, will be able to place 10 tons in LEO for $20 million ($2,200 per kilogram) and 5 tons into GTO for $20 million ($4,400 per kilogram).[16] Whether such low launch costs are possible remains to be demonstrated.

The miniaturization of satellite technology also permits cost savings. A satellite of a given size can perform more missions, or the same missions can be done with smaller satellites. The latter approach would allow the use of much smaller launch vehicles, which a number of developers believe would reduce launch costs.[17] For example, the SpaceX Falcon I booster under development has a launch mass of only 30 tons, which is much smaller than the vehicles listed in Table 8.1. The goal is to launch satellites of 600 to 700 kg to LEO for $6 million (corresponding to about $10,000 per kilogram). It is also intended to provide rapid response and be ready to launch in 24 hours.[18] Similarly, the Microcosm Sprite vehicle is

14. See, for example, Peter Taylor, "Why Are Launch Costs So High?" September 2004, http://www.ghg.net/redflame/launch.htm, accessed January 21, 2005.

15. William Scott, "Rapid Response," Aviation Week and Space Technology, April 7, 2003, 66.

16. Covault.

17. See, for example, Matt Bille and Robyn Kane, "Practical Microsat Launch Systems: Economics and Technology," Paper SSCO3-III-3, AIAA/USU Conference on Small Satellites, August 2003, http://www.mitre.org/work/tech_papers/tech_papers_03/kane_mls/kane_mls.pdf, accessed January 21, 2005.

18. SpaceX, http://www.spacex.com, accessed January 21, 2005; Covaut; Michael Dornheim, "Quick, Cheap Launch," Aviation Week and Space Technology, April 7, 2003, 70.

designed to place a 300 kg satellite in LEO for $1.8 million ($6,000 per kg).[19] A current goal for the RASCAL launcher is to place a 75-kg payload in orbit for $750,000 ($10,000 per kg).[20]

The Orbital Sciences Minotaur and air-launched Pegasus vehicles currently launch small payloads, but at a cost of $40,000 to $50,000 per kilogram— significantly higher than the costs projected for the Falcon I and Microcosm Sprite.[21]

In addition, sufficiently small satellites can be small enough to piggyback on another satellite's launch, often leading to substantially reduced launch costs.

19. Dornheim.

20. Leonard David, "Military Space: Securing the High Ground," Space.com, April 2, 2003, http://www.space.com/businesstechnology/technology/higher_ground_030402.html, accessed January

21. 2005. 21. Minotaur is reported to place over 400 kg in LEO at a cost of up to $19 million (Bille and Kane). Pegasus is reported to cost $22–26 million to place 500 kg in LEO (Dornheim).

Appendix - 8A

Potential and Kinetic Energy of Satellites

The potential energy of a satellite is a measure of the energy required to lift it to its orbital altitude, whereas the kinetic energy reflects the amount of energy required to give the satellite its orbital speed. For a circular orbit at altitude h, the kinetic energy of a mass m due to its orbital speed is

$$KE = \frac{1}{2} m V^2 = \frac{GM_e m}{2r} \tag{8.1}$$

where r = h + Re and the second equality uses Equation 4.2 for the orbital speed.

It is useful to discuss the potential energy in two ways. The first sets the zero of potential energy at the Earth's surface, since this is useful in comparing how much kinetic versus potential energy a satellite gains by being placed into orbit.

With this choice, the potential energy of a mass m at an altitude h is

$$PE = GM_e m \left(\frac{1}{\mathrm{R}_e} - \frac{1}{R_e + h} \right) = GM_e m \frac{h}{R_e r} \tag{8.2}$$

The ratio of potential to kinetic energy of a mass in a circular orbit is then $2h/R_e$. Thus, kinetic energy dominates potential energy out to $h = R_e/2$ = 3,185 km. The total energy of a mass in circular orbit at altitude h is

$$PE = KE = \frac{GM_e m}{2R_e} \left(\frac{2\frac{h}{R_e} + 1}{\frac{h}{R_e} + 1} \right) = \frac{1}{2} m V^2 escape \left(\frac{1 + \frac{h}{r}}{2} \right) \tag{8.3}$$

where V_{escape} is the escape velocity. The last expression shows that the energy is bounded by the kinetic energy the object would have if its speed was equal to the escape velocity. If instead the zero of potential energy is set at infinite distance from the Earth (i.e., outside the gravitational well of the Earth), the potential energy is given by

$$PE = -\frac{GM_e m}{r} = -2\,KE \tag{8.4}$$

for all r, which is a special case of the Virial Theorem.

Appendix - 8B

The "1/2 Rule"

The "1/2 Rule" states that a ballistic missile that can carry a given payload to a maximum range R on the Earth can lift that same mass to an altitude of roughly R/2 when launched vertically.

For the case of a missile launched on a flat earth, it is straightforward to show that the 1/2 Rule is exact. As a result, it holds for short-range missiles, for which the curvature of the Earth can be neglected.

In the flat-earth approximation, the gravitational acceleration is constant with altitude. Consider a missile of mass m fired vertically with a speed V. The maximum height h it reaches is found by equating kinetic and potential energy:

$$\frac{1}{2}mV^2 = mgh \Rightarrow h = \frac{V^2}{2g} \tag{8.5}$$

The time it takes to reach its apogee is

$$h = \frac{1}{2}gt_{apogee}^2 \Rightarrow t_{apogee} = \frac{V}{g} \tag{8.6}$$

To maximize missile range on a flat earth, the missile is launched at 45°. If the initial missile velocity is V, the vertical and horizontal components (Vv and Vh) are both equal to $\left(V/\sqrt{2}\right)$. The missile's range is then given by the (constant) horizontal speed multiplied by the time it takes the missile to climb to apogee and fall back to Earth:

$$range = 2t_{apogee}V_h = 2\left(\frac{V_v}{g}\right)V_h = \frac{V^2}{g} \tag{8.7}$$

which is twice the maximum height found above. For the round-earth case, the gravitational acceleration varies with altitude. The maximum altitude a missile can reach when fired vertically is estimated by again setting the potential energy at the maximum altitude h equal to its initial kinetic energy. Assuming the missile has a speed V at the Earth's surface, then

$$GM_e m\left(\frac{1}{R_e}-\frac{1}{R_e+h}\right)=\frac{1}{2}mV^2 \tag{8.8}$$

Solving for the maximum altitude h gives

$$h=R_e\left(\frac{(V/V_0)^2}{2-(V/V_0)^2}\right) \tag{8.9}$$

where

$$V_0\equiv\sqrt{\frac{GM_e}{R_e}}=7.91\text{ km/s} \tag{8.10}$$

is the orbital speed of a circular orbit with a radius equal to R_e. This equation shows that $V=\sqrt{2}V_0$ = 11.2 km/s gives $h=\infty$ and is therefore the escape velocity from the Earth. It also shows that V = V_0 gives h = R_e = 6370 km.

For V = 3 km/s, which corresponds to a 1,000-km range ballistic missile, this equation gives h = 0.078R_e = 495 km, and for V = 7.2 km/s, which corresponds to a 10,000-km range ballistic missile, this equation gives h = 0.71R_e = 4525 km. These results show that the 1/2 Rule holds for missiles with range small compared to the radius of the Earth and continues to hold approximately even for longer ranges.

Section - 9

Space-Basing

The ability of satellites to orbit over any part of the Earth has led military planners to consider expanding beyond the current military uses of space. Satellites could be equipped to attack targets on the Earth, to intercept ballistic missiles, to defend U.S. satellites, and to inspect and/or attack enemy satellites.

However, technical factors will determine whether satellites make sense for a particular mission. Examples include how time sensitive a mission is, and the corresponding responsiveness required of the satellite system; how expensive it is to accomplish the mission from space; and what alternate means exist for carrying out the mission.

This section considers the implications of the technical issues discussed in the previous sections for space-based ground attack weapons, space-based boost-phase missile defense, and the military space plane, which has been envisioned for a range of missions. Section 12 considers space-based ASATs and compares them with ground-based anti-satellite weapons (ASATs).

The space-based laser (SBL) is another system that has been discussed for defending against ballistic missiles, attacking satellites, and attacking air and ground targets.[1] Its key attraction is that laser beams travel at the speed of light, so the time to deliver an attack would be set by the time required to position the beam

1. An analysis of space-based lasers is given in Bob Preston, Dana Johnson, Sean Edwards, Michael Miller, and Calvin Shipbaugh, Space Weapons, Earth Wars (Santa Monica, CA: RAND, 2002), 24-35. See also, Bruce DeBlois, Richard Garwin, R. Scott Kemp, and Jeremy Marwell, "Space Weapons: Crossing the U.S. Rubicon," International Security 29 (2004): 1-34.

and for the beam to dwell on the target. In addition, the speed of the beam could allow the lasers to be placed in high-altitude orbits (assuming they had sufficiently high power and accurate control over the beam direction), thus reducing the number needed for global coverage of the Earth and all satellites in low orbits. However, the technology for a usable SBL does not currently exist and will not for the foreseeable future.[2] For this reason, this report does not consider it.

Space - Based Kinetic Ground -Attack Weapons

Placing ground-attack weapons in orbit would in principle allow a country to attack any point on Earth. The satellites in the constellation could carry a variety of conventional weapons,[3] including high explosives and kinetic-energy weapons that would use the energy from their high speed to attempt to destroy targets by smashing into them.

Proponents of these systems are interested in a fairly rapid response time, requiring the satellites to be based in low earth orbits. Exactly how many orbiting weapons would be required depends on the desired response time. The system considered below has the capability to deliver weapons to any point on Earth within 30 to 45 minutes of a decision to do so, since that is comparable to the flight time of a long-range ballistic missile, which could also deliver such an attack.

This section compares the costs of delivering ground-attack conventional weapons from space to delivering them by ballistic missile. There are important technical issues related to the transit of weapons through the atmosphere at high speed, including guiding them accurately and dealing with the intense heating, but these issues will be similar for both basing modes. While longrange ballistic missiles have in the past been restricted to

2. In response to Congress slashing funds for the space-based laser program, the Missile Defense Agency disbanded the program, although some technology development continues as part of other programs. See Laura Colarusso "Space-Based Laser Program Office Dismantled, Tech Demo on Hold," Inside Missile Defense 8 (November 13, 2002): 7.

3. The Outer Space Treaty of 1967, which has been ratified by the great majority of countries, forbids the stationing of weapons of mass destruction in space (http://www.oosa. unvienna. org/SpaceLaw/outersptxt.htm, accessed February 2, 2005).

nonconventional roles such as nuclear and possibly biological warheads, there is currently some interest in using them for conventional roles, which would allow them to carry out the same kinds of missions discussed for space-based weapons.[4]

The simple model presented below illustrates the factors that determine the number of satellites required for the desired response time, and allows this system to be compared with delivery of weapons by ballistic missiles. A more detailed calculation is needed to look at tradeoffs between numbers of satellites, deployment altitude, and mass.

Constellation Size

The response time for a space-based ground attack weapon, which is the time from a decision to launch the attack until the weapon hits the target, consists of two parts: the time required for the satellite to get into position so it can de-orbit the weapon toward the target, and the time required for the weapon to de-orbit and reach the ground once the satellite is in position.

The number of satellites required depends on the desired response time, the portion of the Earth to be covered, and the lateral reach of the satellites (the distance each satellite can travel perpendicular to its ground track to strike a target). For example, consider a single satellite in a polar orbit at 500- kilometer altitude that has the propellant needed to allow it to de-orbit and attack a ground target. This satellite orbits the Earth in just over 90 minutes. After one orbit, its ground track crosses the equator 2,600 kilometers west of where it crossed on the previous orbit, due to the rotation of the Earth.

For simplicity, initially assume the satellite is able to reach out laterally as it descends toward the Earth and attack ground targets up to 1,300 kilometers to either side of its ground track.[5] It would then be able to attack any point within a 2,600-kilometer-wide band centered on its ground track. As a result, adjacent bands of ground

4. Important political considerations must be taken into account in when considering such a mission, but this report does not discuss them.

5. Lateral reach can be achieved by giving the object some component of ΔV perpendicular to the plane of the orbit and by designing the object to use the strong aerodynamic forces as it reenters to move it in a lateral direction.

coverage would just abut one another at the equator (see Figure 9.1), and the collection of bands would cover the Earth's surface in 12 hours (during each orbit, the satellite crosses the equator twice, once while going up towards the north pole, and once coming down). The bands will overlap and give greater coverage near the poles. This one satellite in a 500-km-altitude polar orbit would be able to attack any point on Earth twice a day.

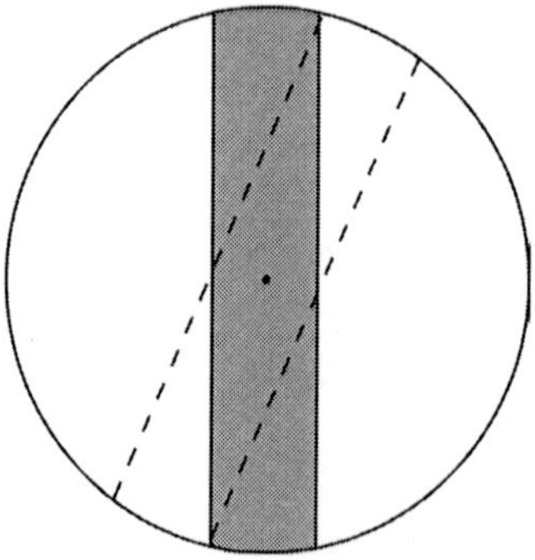

Figure 9.1. This figure shows a view looking down at the north pole of the 2,600-km-wide bands surrounding two successive ground tracks (one white, one shaded) of a satellite in a 500 km altitude polar orbit. Note that the two bands abut at the equator, which is the edge of the circle.

If the time requirement was to have one satellite in position to attack any point on Earth within 12 hours of a decision to attack, this mission could be accomplished with one satellite in orbit.[6] One way to shorten the time required for an attack would be to add satellites in other orbital planes. For example, adding a satellite in a polar orbit rotated by 90° to the first would cut in half the maximum time to deliver an attack—to 6 hours.

Adding more satellites in other orbital planes would further reduce the time. Since 8 bands of the type described above cover the Earth at the equator, placing satellites with a lateral reach of 1,300 km in eight equally spaced orbital planes would ensure that a satellite was in a position to attack any target within one orbital period, about 90 minutes.

6. On average, this system would be able to attack within about six hours, but could not guarantee an attack in less than twelve.

Assuming a lateral reach of 1,300 km for the satellite simplified the above analysis, but would require very large lateral speeds for the de-orbit times considered here.[7] A more realistic value for the lateral reach may be only a few hundred kilometers. The discussion below assumes a lateral reach of 650 km, which would double the number of orbital planes needed to get a satellite in position to attack in 90 minutes, resulting in 16 planes.

Once the satellite is in the proper position, it would accelerate the weapon out of its orbit toward the ground. As discussed in Section 6, a de-orbiting time of 10 to 15 minutes from an altitude of 500 km can be achieved with a ΔV of 0.7 to 1.0 km/s. So this configuration—16 orbital planes with one satellite per plane—would allow global attacks anywhere on Earth within 100 to 110 minutes.

The attack time could be further reduced by placing several satellites in each orbital plane. For example, placing six equally spaced satellites in each of the 16 orbital planes would reduce the maximum time for a satellite to move into position to 15 minutes instead of 90. Adding the de-orbit time of 10 to 15 minutes gives a total attack time of 25 to 30 minutes.

Such a constellation—which could attack any point on the Earth within about 30 minutes—would comprise 96 satellites (16 planes × 6 satellites per plane), and would therefore have an absentee ratio of 96. If the requirement was having two satellites in position to attack one or two targets in the same region at any time, the constellation would need to be doubled to 192 satellites.

If the constellation instead consisted of three satellites in each plane, for a total of 48 satellites, its response time would be about 45 minutes.

---

7. Once the satellite de-orbited to a low enough altitude it could use atmospheric forces to turn and move in a lateral direction. However, in the case considered here, the de-orbit time from 80 km altitude to the ground would be roughly 200 seconds. Reaching laterally 1300 km during that time would require an average lateral speed of 1,300 km/200 s = 6.5 km/s, which is higher than the average speed of the re-entering satellite at altitudes below 80 km. Moreover, attempting to turn quickly may lead to unacceptably high forces that could damage the satellite. The satellite could also use propellant to give it a lateral speed at high altitude, in addition to the vertical ΔV used to de-orbit. A lateral ΔV of 1 km/s would result in a lateral reach of 600 to 900 km during a de-orbit time of 600 to 900 seconds, but the propellant required would add 40% to the mass of the satellite.

Thus the responsiveness required of the system quickly drives up the size of the constellation, as does the number of satellites required to be in position at any time. As noted above, increasing the lateral reach of the satellites would reduce the number of satellites needed. Conversely, a smaller lateral reach than assumed here would increase the number of required satellites.

The constellations considered so far assumed the satellites were in polar orbits. These orbits give complete coverage at the equator, but provide overlap of coverage at mid and high latitudes; in other words, a constellation that covered the equator a minimum of twice a day would cover areas near the poles many more times a day.

If the attacker were willing to give up coverage of the polar regions, it could reduce the number of orbital planes required by using orbits with inclinations less than 90°. Recall that the ground track of a satellite in an orbit with inclination θ does not reach beyond a latitude of θ (see Figure 5.1). The inclination of the orbits would therefore need to be approximately as large as the latitude of important potential targets. In addition, if the attacker was willing to have some gaps between the bands near the equator, giving a somewhat longer response time there, coverage could be optimized for targets at midlatitudes.

For a satellite with a lateral reach of 650 kilometers, seven orbital planes with inclination of 45° will cover that part of the Earth with latitude between about 50° north and 50° south,[8] with the coverage optimized for roughly 30° to 50° latitude (both above and below the equator; see Figure 9.2). These latitudes include the Middle East; North Korea; most of Europe, the United States, and China; and part of Africa and South America. They exclude essentially all of Russia.

A constellation of 42 satellites, with 6 in each of the 7 orbital planes, will provide single-satellite coverage of the region between about 30° and 50° with an attack time of 25 to 30 minutes. This constellation would have no coverage above about 55° and would require somewhat longer times on average to attack targets below 30°. To be able to attack more than one (say, n) targets in the same

8. As noted in Section 4, the ground track of the satellite will only reach to 45°, but the lateral reach of the satellite allows it to reach somewhat higher latitudes.

region at the same time with this responsiveness, the constellation would need 42× n satellites.

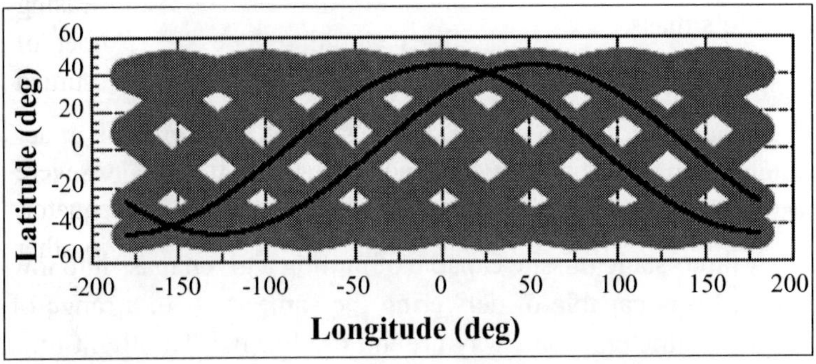

Figure 9.2. The figure shows the ground coverage (gray areas) of satellites in seven equally spaced orbital planes with inclination of 45°, assuming the satellites can reach laterally 650 km as they de-orbit. The two dark lines are the ground tracks of two of the satellites in neighboring planes. This constellation can provide complete ground coverage for areas between about 30° and 50° latitude (both north and south), less coverage below 30°, and no coverage above about 55°. Due to the rotation of the Earth, satellites would pass over the holes in the coverage shown above, but it would on average take longer to attack targets in these areas.

If the response time were relaxed to 45 minutes, a similar 7 orbital plane constellation could use just three satellites per orbit, for a total of 21 satellites, to provide single-satellite coverage of the same regions.

For all the constellations discussed above, the ΔV available for de-orbiting was 0.7 to 1.0 km/s. If the satellite is designed to provide a much higher ΔV to the weapon, thereby increasing its de-orbiting speed, a smaller constellation could provide the same response time by placing the satellites in orbits at higher altitudes than considered above.[9] A larger ΔV would also increase the speed at which the kinetic weapon hit the Earth, resulting in greater destructive power. However, as Section 6 discusses, the atmospheric drag and heating increase rapidly with the speed of a re-entering object, placing practical limits on the

9. Wang Ting, Beihang University (Beijing University of Aeronautics and Astronautics), personal communication, April 2004.

speeds that could be used. Moreover, as discussed in Section 7, increasing the ΔV available to the satellite, including for extending the weapon's lateral reach, would rapidly increase the satellite's mass.

Comparison to Delivery by Ballistic Missile

The launch requirements of these space-based ground attack systems can now be compared with those of a ballistic missile system that provides similar capability.

A three-stage missile capable of putting a given mass into low earth orbit is capable of delivering the same mass to a range of 20,000 kilometers—half way around the Earth.[10] The flight time would be roughly 45 minutes. This one ballistic missile could therefore provide global coverage with the same response time as the constellation described above of 48 satellites with three satellites in each of 16 orbital planes.

However, for the space-based system, part of the mass placed into orbit will need to be devoted to propellant to de-orbit the weapon. As noted above, our calculations show that accelerating a satellite out of a 500-kilometer altitude orbit so that it will reach the ground in 10 to 15 minutes would require a ΔV of 0.7 to 1 km/s. For a satellite to achieve this ΔV, it must carry an additional 25 to 40% of its mass in propellant.[11] Thus, the weapon itself would constitute only 60 to 75% of the mass in orbit. Designing a system with higher ΔV to give the weapon a much higher speed as it de-orbited (to reduce the de-orbit time or increase the lateral reach) would increase the propellant mass. For example, for a satellite carrying propellant for a ΔV of 5 km/s, the weapon would constitute less than 20% of the mass in orbit, with 80% being the propellant for de-orbiting.

This information can be used to directly compare the launch requirements of a space-based and ground-based system with a 45-minute response time. One ground-based ballistic missile

10. Such missiles could also attack targets at shorter ranges by deliberately wasting fuel.

11. This calculation assumes conventional technology thrusters, with an exhaust velocity of 3 km/s.

capable of placing a mass m in orbit could deliver a weapon of mass m to a target anywhere on Earth within 45 minutes. For the 48-satellite constellation, the capability to deliver mass m to any point on Earth within 45 minutes would require a total mass of 60m to 67m in orbit (48 satellites, each with a weapon of mass m and propellant mass 0.25m to 0.4m for de-orbiting). Placing this mass in orbit would require 60 to 67 launches by missiles of the type used by the ground-based system. The 21-satellite constellation considered above that does not cover areas above 50° latitude would require a total mass of 26m to 29m in orbit to be able to deliver a mass m to a target in this region within 45 minutes.

This is actually an underestimate of the mass required in orbit since satellites in low earth orbits would require additional propellant for stationkeeping. (Placing the satellites in higher orbits, where there is less atmospheric drag, would reduce the need for stationkeeping fuel, but would require more propellant for de-orbiting to meet the same time goal.) The satellites would also carry components that the missile-launched weapon would not—such as tanks to hold the propellant, solar panels, and ground communications systems— and these will also add mass.

This analysis, based only on launch requirements, shows that acquiring the capability to attack a ground target within 45 minutes would be many tens of times more costly if done from space than from the ground.[12]

Next consider a response time of roughly 30 minutes. Global coverage of one target could be obtained by deploying at widely spaced locations two missiles with a range of 10,000 km and a corresponding flight time of 30 minutes. (However, three or four missiles might be needed, depending on geographic constraints.) To compare the launch capacity required for a spacebased system having a 30-minute response time, we again need to consider the mass of de-orbiting propellant that must be launched as part of the spacebased weapons. But we must also take into account the fact that a missile can deliver only about two-thirds as much mass to

12. In addition, the space-based system would entail significant additional costs, such as building the satellites that deliver the weapons back to Earth.

low earth orbit as it could deliver to a range of 10,000 km.[13] This means that it requires about 1.5 times as much space lift to place a mass m into orbit as it does to deliver it to 10,000 km.

Taking these considerations into account, the 96-satellite system discussed above (global coverage in 30 minutes using 16 planes with 6 satellites in each) would require a total mass of roughly 125m (96 satellites plus their de-orbiting propellant) in orbit to be able to deliver a mass m to a target anywhere on Earth within about 30 minutes. Using a missile that could deliver a mass m to a range of 10,000 km, roughly 190 launches (125× 1.5) would be required to place this mass in orbit. The smaller 42-satellite constellation that would cover the Earth between latitudes of 50° north and 50° south would require a total mass of roughly 55m in orbit, and about 85 missile launches (55× 1.5) to place this mass in orbit. Since two ballistic missiles would be needed to give equal coverage in either case, the smaller 42-satellite constellation would require roughly 85/2 = 42 times as much launch capacity as the ground-based system and the 96-satellite system would require roughly 190/2 = 95 times as much launch capability.[14] As noted above, these are underestimates of the relative launch requirements of a space-based system and a ground-based system of two ballistic missiles; moreover, developing a space-based system would entail additional costs beyond those for launch.

This discussion is most relevant to a country making the choice between building long-range ballistic missiles or placing ground-attack weapons in space. However, the five declared nuclear weapon states already have longrange ballistic missiles based on submarines and on the ground. For those countries,

13. A 10,000-km range missile will typically burn out at an altitude of several hundred kilometers with a speed of roughly 7 km/s. This speed is about 10% too low to place a satellite in a circular orbit at those altitudes. Reaching the necessary speed would require reducing the payload by roughly a third. (Steve Fetter, University of Maryland, personal communication, July 2004. For several different missiles, he calculated the mass that a missile could launch into low earth orbit [200 and 500 km altitude] and compared it with the mass he calculated the same missile could send to 10,000 km range.)

14. Of course, a land-based missile that was out of range of the target would not be launched in an attack, whereas all of the satellites would have to be launched into orbit to put in place the space-based constellation with the desired response time.

deploying ballistic missiles with kinetic ground attack weapons would not require large additional investments.

Space - Based Boost- Phase Missile Defense

Boost phase missile defense systems would be designed to destroy a missile during its boost phase—when the rocket is still burning. For long-range missiles, the boost phase lasts for only 3 to 4 minutes, requiring a defense with a very short response time.[15]

The United States is considering or developing several types of Earth-based boost phase systems, including ground- or ship-based interceptors with kill vehicles that would use their sensors to home on the bright flame of the boosting missile and attempt to destroy the missile by direct impact. These interceptors would need to be based close to the missile launch site to reach the missile during its boost phase. Another option under development is air-based lasers, which must be within a few hundred kilometers of their target to destroy it.

A space-based boost phase defense would consist of a constellation of spacebased interceptors (SBI) in low earth orbit. These satellites would remain in orbit until a missile launch was detected; an SBI near the missile launch site would then use its onboard propulsion to accelerate out of orbit and maneuver toward the missile. The intercept must occur above 80 to 100 km altitude, since the interceptors are not designed to operate lower in the atmosphere where they would be subject to high heating. As a result, the defense would be unable to defend against shorter range missiles, since these would burn out too low in the atmosphere to be engaged by the space-based interceptors.

Proponents of deploying a space-based system argue that it could defend against missiles launched from anywhere in the world. Indeed, the geographic and political restrictions on where surface- or air-based defenses could be located means that space-based interceptors may be uniquely able to reach missiles launched from some locations during their boost phase. Thus, unlike the case for ground-attack weapons, it is not possible to make a direct comparison between space-based and Earth-based

15. Shorter range missiles would have shorter burn times.

systems that can carry out the same mission. Instead, we assess the ability of the space-based system to perform its mission and the launch capacity required to place the system in space. We end by discussing the ASAT capability of SBIs.

Vulnerabilities of SBI

A key technical difficulty of a space-based missile defense is the vulnerability of the system to attack. The SBI could be tracked from the ground and their locations would be well known. Because the SBI would be in low-altitude orbits (300 to 500 km), they could be attacked by ASATs on short-range missiles with ranges of 600 to 1,000 km. Such missiles would burn out too low for the SBI to intercept them in their boost phase. If the SBIs were programmed to ignore short-range missiles, the SBI would be vulnerable to attack while in their orbits. But because an SBI must be launched quickly after detection of a missile launch, the SBI might have to launch before it could determine the range of the missile. Causing an SBI to be launched would remove it from orbit and deplete the constellation as effectively as destroying it with an ASAT.

Since short-range missiles are much less expensive than long-range missiles, a country could launch enough ASATs on short-range missiles to create a hole in the constellation. The attacking country could launch a long-range missile through this hole when it reappeared after an orbital period of roughly 90 minutes or could even plan to launch from a location the hole passed over shortly afterward.

Many systems containing satellites can be structured so that the vulnerability of individual satellites does not cause the overall system to fail to complete its mission. However, space-based missile defense is an exception. Because only the SBI closest to the region where a missile is launched are able to engage the missile in the time available, destroying interceptors and creating a hole in the system prevents the defense from engaging missiles launched through that hole.

While no countries currently deploy ASATs, countries that have developed the technical sophistication and the aerospace expertise to launch long-range missiles would also be expected to have the technical capability to build ASATs that could attack SBIs. Developing or acquiring the capability to carry out such attacks would become a high priority for any country that had developed long-range missiles.

Even if the SBIs were not attacked, the defense could be easily overwhelmed, although this may not be the most cost effective way of foiling a space-based system. As discussed below, SBIs have a large absentee ratio, i.e., a large number of satellites are needed in a constellation to ensure that even one SBI is in position to engage a missile launched from a particular location. The total number of SBI needed in the constellation in order to engage n simultaneous launches from the same location is n times the absentee ratio. To avoid large numbers of SBIs in orbit, most proposals for space-based missile defense consider systems that would be able to engage only a small number of missiles (n is typically one or two) launched nearly simultaneously from the same location. If the attacker launched more than that number of missiles, the defense would not be able to engage them all.

While a ground-based defense system can also be overwhelmed by simultaneous launches, the large absentee ratio of the space-based system means that it is much more expensive to increase the number of simultaneous launches it can handle compared with a ground-based system.

Constellation Size and Launch Requirements

A space-based missile defense would require large numbers of interceptors and deploying even a thin defense would be expensive. Recall that a groundattack system with a 30-minute response time and global coverage would require nearly 100 satellites. A missile defense with a response time of only a few minutes would require many hundreds of satellites, as discussed below.

Similar to the ground attack system analyzed above, the structure of the missile defense constellation would depend on what parts of the Earth the system was intended to cover. Truly global coverage would require some satellites in polar orbits. A system using satellites in orbits with inclinations less than about 45° would not be able to defend against launches from locations with latitudes above about 45°. Such a system would cover the Middle East and almost all of the United States and China, but would not cover Russia or northern Europe.

The Brilliant Pebbles system proposed as part of the Global Protection Against Limited Strikes (GPALS) system in the early 1990s by the first Bush administration was intended to include 1,000 SBIs for global coverage of one or two missiles launched simultaneously from a single site.

A technical analysis of boost-phase missile defense published by the American Physical Society (APS) in July 2003 found that a similar number of interceptors were required.[16] The APS panel considered a constellation of SBIs in orbits at an altitude of 300 km that would place a minimum of one and occasionally two interceptors within range of any launch site between 30° and 45° latitude (which includes North Korea and the Middle East), but would provide no coverage above 45° and somewhat limited coverage near the equator. They determined that this system would require roughly 1,600 SBIs to engage solid fueled missiles (with a boost phase of 170 seconds), and roughly 700 SBIs to engage liquid-fueled missiles (with a boost phase of 240 seconds).

Increasing the regions of the Earth covered by the system would significantly increase the number of SBIs needed; global coverage would roughly double the number required.[17] Moreover, a system that could engage a minimum of two missiles launched simultaneously from the same area or that could launch two interceptors at one attacking missile would require doubling the number of satellites.

---

16. American Physical Society (APS), Report of the American Physical Society Study Group on Boost-Phase Intercept Systems for National Missile Defense, July 2003, http://www.aps.org/public_ affairs/popa/reports/nmd03.html, accessed January 5, 2005.

17. Conversely, by restricting the coverage of the system to considerably smaller geographical areas, the system can be designed to have many fewer SBIs. One recent study looks at concentrating the coverage in a band of latitude that is less than 400 km wide, which is only designed to engage missiles from a restricted set of launch sites (G. Canavan, "Estimates of Performance and Cost for Boost Phase Intercept," September 24, 2004, www.marshall.org/pdf/materials/ 262.pdf , accessed December 20, 2004; I. Oelrich and S. Fetter, "Not So Fast," February 1,2005, www.fas.org/resource/01312005123203.pdf, accessed February 15, 2005). While North Korea is discussed as a geographically small country for which such a system might be appropriate, the size and location of North Korea makes sea-based boost phase systems appear feasible, so that the space-based system should be compared with surface-based alternatives for this case.

The APS study also showed that a substantial ΔV would be required to give the SBI sufficient maneuvering capability to be effective in the face of intrinsic uncertainties in the engagement. The required ΔV has two main contributions: that needed to accelerate the interceptor out of its orbit and toward the boosting missile and that needed for maneuvering to achieve an intercept. The study determined that the SBI would require a thruster that could provide a ΔV of 4 km/s with high acceleration to kick it out of orbit with sufficient speed to reach the boosting missile in time. Reducing this speed would require the SBIs to be spaced more closely in orbit, increasing the size of the constellation. The study also determined that an additional ΔV of 2.5 km/s would be required to allow the kill vehicle to maneuver and home on the boosting missile, which is an accelerating target that could be deceptively maneuvering.[18]

As discussed in Section 7, large values of ΔV require large propellant masses. Assuming technology available in the next decade, the APS panel determined that the engines and propellant would increase the mass from about 60 kg for the unfueled kill vehicle to more than 800 kilograms per SBI.[19]

Since the SBI is intended to remain in orbit for many years at a low altitude, it must also carry propellant for stationkeeping. In addition, typical designs assume that while in orbit the SBI would be housed in a garage or life jacket to protect it from radiation and debris and that would have communications equipment and solar panels for power—all of which would be left behind when the interceptor was accelerated toward a missile. According to the APS study, adding the garage mass would bring the total mass in orbit for each SBI to more than a ton.

Using these assumptions, a constellation of 1,600 SBI needed to defend against solid-fuel missiles would require a total mass in

---

18. APS, 110.

19. APS, 111. The mass of the (unfueled) kill vehicle for the model used in the APS study is 60 kg, and it carries nearly 80 kg of propellant for maneuvering as it attempts to strike the missile. The additional mass is due to the thruster and propellant used to accelerate it out of orbit.

orbit of nearly 2,000 tons.[20] Assuming a launch cost of $20,000 per kilogram (see Section 8), the launch cost would be roughly $40 billion. A constellation of 700 SBI needed to defend against liquid-fuel missiles would require a total mass in orbit of 850 tons, leading to an estimated launch cost of $17 billion. Recall that these systems could engage only one or two missiles launched simultaneously from a single site. Placing 1,000 tons in orbit would require the equivalent of more than 100 Delta or Atlas II/III launches or more than 50 Atlas V launches. In recent years, the launch rate has been roughly seven Delta launches and five Atlas launches (a mix of II, III, and V),[21] so launching such a defense would require a significant increase in launch capability.

Moreover, assuming the satellites have a 10-year lifetime, roughly 100 satellites would need to be launched every year to maintain a 1,000-satellite constellation. This would entail a cost of $2 billion per year, given a launch cost of $20,000 per kilogram.

These total mass figures could be decreased by reducing either the number of interceptors or the mass of the SBI. The issue of how light the SBI can be made is controversial and depends, in part, on the timeline considered. The APS study based its model of the SBI on the technologies it judged to be realistic in the next decade. It considered further possible reductions in the mass of the SBI that might reduce it by about 60%.[22] Other estimates have raised the possibility of even lighter SBIs—considerably lighter than the lightest APS model—although the timeline and other details of these estimates have not been made public.[23]

---

20. For the case of defending against solid-fuel missiles, the APS study finds a total mass of 2,000 tons (APS, 114, 126).

21. See, for example, "History of the Delta Launch Vehicle," http://kevinforsyth.net/delta/, accessed January 5, 2005, and "Atlas Launch Information," http://home.cfl.rr.com/atlas/ info_ atlas.html, accessed January 5, 2005.

22. APS, 125-126.

23. For example, in its July 2004 study of boost-phase defenses, the Congressional Budget Office (CBO) considered the effects of using a model for the kill vehicle that was briefed to the CBO by researchers at the Lawrence Livermore National Laboratory in November 2003 (CBO, Alternatives for Boost-Phase Missile Defense [July 2004], 24). The model used for the kill vehicle has an unfueled mass of 11 kg (Canavan, 6) and a fueled mass of 30 kg; the SBI can accelerate out of orbit using a ΔV of 6 km/s and has a mass of 442 kg, without the garage (CBO, xvii). The CBO report states that producing this model "would require a technological leap in miniaturization" (CBO, 43).

Reducing the SBI mass may make it possible to increase its ΔV without a prohibitive increase in propellant mass, and this increase in speed may increase the optimal orbital altitude and decrease the number of interceptors required in the constellation. For example, a July 2004 Congressional Budget Office analysis considered a fast, lightweight SBI having a ΔV of 6 km/s rather than the 4 km/s assumed by APS, a fueled kill vehicle mass of 30 kg rather than the 136 kg assumed by APS, and a garage mass of 90 kg rather than the 440 kg assumed by APS. CBO found that if such an SBI could be built, it would reduce the total number of interceptors in orbit compared with the APS values by roughly a factor of three (for defending against solid-fuel missiles) to 4.5 (for defending against liquid-fuel missiles). In addition, it would reduce the total mass of interceptors in orbit compared with the APS values by roughly a factor of seven (for defending against solid-fuel missiles) to ten (for defending against liquid-fuel missiles).[24]

ASAT Capability of SBI

While the large constellation of SBIs needed for missile defense could not be deployed for many years, small numbers of prototypes could be deployed earlier. These systems are important to consider since they could have the capability to attack satellites with little warning, including satellites in geosynchronous or semisynchronous orbits.

Assuming it was designed with sensors that could detect a satellite in orbit, an SBI designed to intercept a boosting missile would have more than enough maneuverability to intercept a satellite in orbit. Moreover, the large ΔV the SBI would possess for accelerating out of orbit would also allow it to change its orbit to attack satellites in orbits significantly different from its own, including geosynchronous orbit.

The orbital speed of the SBI would be roughly 8 km/s; adding the 4 km/s it would need to reach a boosting missile, it could reach a total speed of up to 12 km/s.

---

24. CBO, 30, 35. The SBI model called Option 4 in the CBO report has a higher average acceleration than the APS model, so the number of interceptors required is less than that calculated by APS.

Our calculations show that such a speed would allow it to travel from low earth orbit to geosynchronous orbit in an hour and a half and still have a speed of nearly 7 km/s at that altitude. Ground observations could determine the location of the satellite to be attacked with sufficient accuracy to launch the interceptor and allow the onboard sensors to detect the satellite when it was close enough.

Whether a kill vehicle designed solely for missile defense could be used to attack satellites in this way depends on details of its design, such as the type of sensors it contains and the length of time it is designed to operate (a matter of minutes to reach a boosting missile versus an hour to reach geosynchronous orbit). It is clear, however, that these are design decisions and that these capabilities could be built into the SBI to give them the capability to also serve as high-altitude ASATs. The sensors that are designed to enable the SBI to detect the missile plume during the boost phase may not be suitable for detecting a satellite, but lightweight sensors exist that could be added for the ASAT mission. Since geosynchronous satellites are in the sunlight during all or nearly all of their orbit, they would reflect sunlight and would have a relatively high surface temperature, both of which could be used for homing.[25]

The Military Space Plane

An analysis of maneuverability is especially relevant for the proposed Military Space Plane (MSP), which usually refers to the combination of a maneuvering vehicle that is placed in orbit and the launcher that would place it in orbit. The technical issues raised by these two components differ substantially.

The various missions that have been discussed for the MSP are to launch weapons for prompt, global, precision strike;[26] to

25. For a discussion of the temperature of objects in space, see Appendix A of Andrew Sessler et al., Countermeasures, Union of Concerned Scientists and MIT Security Studies Program, April 2000, http://www.ucsusa.org/global_security/missile_defense/page.cfm?pageID=581, accessed January 5, 2005.

26. For example, one proposal envisions "100-minute or less on-alert delivery time" from bases in the continental United States "to points around the globe" (ONE Team, "The Military Space Plane: Providing Transformational And Responsive Global Precision Striking Power," January 2002, http://www.spaceref.com/news/viewsr.html?pid=4523, accessed December 20, 2004).

carry sensors that could provide reconnaissance in a manner that an adversary could not predict; to launch satellites to either augment or reconstitute space assets; to take part in various types of anti-satellite missions; and to inspect or service satellites that are in orbit. It would not be designed to carry humans. For these missions, the design of the launch vehicle is not particularly important, and existing space-launch vehicles could be used.

Another mission commonly given for the MSP is affordable, rapid, ondemand space-launch capability that could place an object in low earth orbit within 5 to 12 hours of a decision to launch. However, new launcher technology would be required to achieve this goal. While the development of reusable and single-stage-to-orbit launch technologies are often discussed as part of the space plane, they are really part of a broader effort to develop new launch technologies.

This section focuses on the orbiting component of the MSP. This object, often called the Space Maneuver Vehicle (SMV), can be thought of as a small, unmanned Space Shuttle (see Figure 9.3). It would have a cargo bay that could carry a range of different payloads and would carry propellant to allow it to maneuver while in orbit. To reduce costs, it is intended to be reusable, so it would carry propellant to allow it to return to Earth from orbit and land like an airplane.

The SMV could be launched into orbit on an existing launch vehicle, so it can be considered separately from the new launch technologies mentioned above. The SMV would be much smaller than the Space Shuttle, with a mass, including propellant, of 5 to 8 tons (compared with 94 tons for the Shuttle) and a payload capability of up to 1 or 2 tons (compared with roughly 20 tons for the Shuttle). As noted, it would not carry humans. It is expected to draw on technology being developed for a test vehicle planned for flight testing in FY2006 (the X-37 Orbital Vehicle),[27] but the development time of a usable SMV is unknown.

27. X-37 Fact Sheet, http://www1.msfc.nasa.gov/NEWSROOM/background/facts/x37.html, accessed December 20, 2004; B. Iannotta and B. Berger, "NASA Brings Back Plans to Fly X-37 Demonstrator in Orbit for 270 Days, Space News, November 17, 2003, http://www.space.com/ spacenews/archive03/x37arch_120103.html, accessed February 4, 2005.

The term "space plane" suggests that the SMV could be operated like an airplane and could move through space similar to the way an airplane can maneuver in the air. However, this analogy is not appropriate: the physics of orbital dynamics places much greater restrictions on what an orbiting vehicle can do, as Sections 4–6 describe.

Figure 9.3. This photo shows the X-40a, a 6.7 m long model of an SMV built for testing.

(Source: http://boeingmedia.com/ images/ one.cfm?image_id=866, accessed February 7, 2005). ™© *Boeing. Used under license.*

The vision is that the SMV would exploit its maneuverability to carry out the missions discussed above. To deploy multiple satellites into different orbits, the SMV would place itself in the first orbit, release the first satellite, maneuver into the next orbit, release the next satellite, and so on. Using a maneuverable vehicle, or bus, to release several satellites can reduce the number of space launches required to place these satellites in orbit. In fact, a nonreusable bus is used routinely to place multiple satellites in different orbits. Maneuvering a sensor in space is also a mission that requires changing from one orbit to another. Similarly, rendezvousing with different satellites to, for example, inspect, service, or possibly interfere with them, requires placing the SMV in the same orbit as the first satellite, then changing orbits to that of the second satellite, and so on. The propellant required for the SMV to maneuver would place significant limits on the amount of maneuvering it could carry out. The SMV is likely to have a total

ΔV of 3 to 4 km/s in normal operation.[28] As Section 6 discusses, maneuvers within an orbital plane require a ΔV of a few tenths of a kilometer per second (assuming the SMV remains in low earth orbit), but changing orbital planes at low altitudes can require a much greater ΔV. To illustrate this, consider an SMV designed to have a total ΔV of 3 km/s. Some of this total must be used to de-orbit the SMV to bring it back to Earth so it can be reused, leaving roughly 2.5 km/s available for maneuvering. A ΔV of 2.5 km/s would allow the SMV only one plane change of less than 20° at an altitude of 500 kilometers. Even if it were carrying enough propellant to give a total ΔV of 6 km/s, the SMV would only be able to change orbital planes separated by roughly 40°.

As a result, the SMV could deploy several satellites into different orbits within one plane, but would have limited ability to deploy satellites into different planes or to rendezvous with other satellites in different orbital planes. The Appendix to Section 9 contains a detailed discussion of these two missions.

Increasing the maneuverability of the SMV by increasing the ΔV, as Sections 7 and 8 showed, would quickly drive up the SMV mass and the associated launch costs.[29] If the SMV was given a significantly larger maneuvering capability, launching the satellites into orbit individually using multiple separate launches would require less launch capacity.

For example, using an SMV to place three 300-kg satellites into orbits in different orbital planes would, as discussed in the Appendix to Section 9, require launching a total mass of several

---

28. Aaron R. Shirk, "The Feasibility of the Military Space Plane for Rapid Response Surveillance and Reconnaissance Satellite Constellation Deployment," April 17, 1998, http:// www.stormingmedia.us/00/0026/A002653.html, accessed December 15, 2004, discusses several models of an SMV. See also Orbital Sciences, "Re-Energizing America's Space Program," November 21, 2002, http://www.spacecoretech.org/coretech2002/Papers/RLVs/pdfs/ Orbital%20 Presentation.pdf, accessed December 20, 2004, and Air Force Research Laboratory, "Space Maneuver Vehicle," September 2002, http://www.vs.afrl. af.mil/Factsheets/smv.html, accessed December 20, 2004.

29. Shirk suggests that the ΔV available for maneuvering might be increased to 6 km/s by carrying an additional propellant tank. However, the rocket equation shows that the additional propellant required to do so would more than double the mass of the SMV. This additional propellant mass is much greater than the payload of the SMV. Increasing the ΔV of the SMV to 6 km/s therefore seems unlikely.

tons, even if the SMV had only a relatively modest maneuvering capability. In contrast, launching each satellite individually into its own orbit would require putting a total mass of less than a ton into orbit. This approach could be much less expensive since the mass of each satellite by itself is small enough that a small launcher or an air-launched vehicle might be used.

Similarly, the limits on maneuvering have implications for other possible missions for the SMV. For example, an SMV could vary its orbit within its orbital plane to change the revisit time of a sensor it was carrying using only relatively small amounts of ΔV (see Section 6). But it would have limited ability to change the orbital plane of the sensor.

In the same way, the SMV would be able to rendezvous with several satellites in the same orbital plane, but it would have limited capability to rendezvous with satellites in orbital planes with different inclinations. Because of the large mass of an SMV relative to a small satellite that might be specifically designed to inspect other satellites, the SMV would require a much greater propellant mass to carry out the same maneuvers. It would therefore not be well suited to this mission.

This analysis illustrates the significant constraint that the physics of maneuvering places on space systems. These constraints must be taken into account when considering the utility of the space plane in carrying out a particular mission.

It is also important to note that the maneuverability foreseen for the SMV does not represent a new capability: the upper stage of current space launch vehicles provides comparable maneuverability. For example, the Fregat Upper Stage used on the Russian Soyuz launch vehicle has an engine that can be restarted multiple times, and would have a mass of 7.4 tons and a ΔV of more than 4 km/s for a one-ton payload.[30] The main potential advantage of the SMV would appear to be its reusability.

30. Starsem Corp., Soyuz User's Manual, April 2001, 26-27, http://www.starsem.com/services/ images/soyuz_users_manual_190401.pdf, accessed December 20, 2004.

Appendix - 9

Comparison of the Military Space Plane versus Multiple Launches

This appendix considers two potential missions that have been discussed for the military space plane: releasing multiple satellites in orbit and rendezvousing with multiple satellites to inspect, service, or attack them.

Deploying Multiple Satellites

To reduce the number of space launches required to place multiple satellites in orbit, a maneuvering orbital vehicle or bus can be used to place several satellites in different orbits.[1] As noted above, a bus is used routinely to place satellites in different orbits in the same plane, but quickly becomes impractical if the satellites are in different planes. The propellant requirement is an important consideration for deploying a constellation of multiple satellites in multiple orbital planes, which is required for space-based missile defense or groundattack weapons, or a network of communication satellites in low earth orbit.

As an example, consider a Space Maneuver Vehicle (SMV) on a mission to release three identical satellites of mass m_s into three different orbits. Assume the SMV itself has a mass M_{SMV} of 3 tons (without payload or propellant). It must carry a propellant mass

31. Similarly, a maneuvering bus is used to launch multiple nuclear warheads against different targets using a single missile. The bus carries propellant and maneuvers to place each of the Multiple Independently Targeted Re-entry Vehicles (MIRVs) on a different trajectory, each of which is part of an orbit that intersects the Earth. The maneuverability of the bus determines over how large an area the warheads from a single missile can be spread. U.S. MIRV buses can provide a total ΔV of up to about 1 km/s (the START I Treaty limits the total ΔV of a bus to 1 km/s or less).

$m_p^{deorbit}$ of nearly a third of a ton to allow it to de-orbit and return to Earth after releasing the satellites (a ΔV of 0.3 km/s for de-orbiting would require 315 kg of propellant).[2] Assume the SMV is launched into the proper orbit for the first satellite and releases it. It then maneuvers into the orbit of the second satellite, which requires a velocity change ΔV_1 and a corresponding propellant mass mp $^{(1)}$. In doing so, the total mass that must be moved by the thrusters is the mass of the SMV and its de-orbiting propellant, the mass of the remaining two satellites, and the propellant mass $m_p^{(2)}$ needed for the second maneuver to place the third satellite into orbit. The propellant mass $m_p^{(1)}$ needed for this first maneuver is found by using Equation 7.4:

$$m_P^{(1)} = (e^{\Delta V_1 / v_e} - 1)(M_{SMV} + m_P^{deorbit} + 2m_s + m_P^{(2)}) \quad (9.1)$$

where V_e is the exhaust velocity of the engine. Similarly,

$$m_P^{(2)} = (e^{\Delta V_2 / V_e} - 1)(M_{SMV} + m_P^{deorbit} + m_s) \quad (9.2)$$

Using these equations, we can calculate the total launch mass required for specified values of ΔV_1, ΔV_2, and m_s. As an example, consider a case in which the satellites have a mass m_s of 300 kg and for which $\Delta V_1 = \Delta V_2$ ΔV. The results are given in Table 9.1.

Table 9.1. This table shows the total launch mass required to deploy three satellites (each of mass 300 kg) in different orbits, with ΔV needed to maneuver between the release of each of the satellites, as described in the text.

ΔV (km/s)	Total launch mass (metric tons)
0.5	5.7
1.0	7.8
1.5	11
2.0	15
2.5	20

32. These calculations assume conventional thruster technology with a Ve of 3 km/s.

Keep in mind that the total mass of the three satellites in this case is just under one ton. The presence of just the maneuvering vehicle and the propellant for de-orbiting increase the mass that must be launched into orbit to over 4 tons; the additional propellant for maneuvering can further increase the mass by an additional 1.5 to 16 tons. A mass of 8 tons corresponds roughly to the capacity of a Delta III launcher (see Table 8.1); 20 tons is greater than the launch capacity of an Ariane 5 and is the maximum capacity of an Atlas V launcher. Clearly, building in the capability to place multiple satellites into different orbits adds significantly to the launch requirements: if significant maneuverability is required, it will be cheaper to launch the satellites on separate launchers.

To put the numbers in Table 9.1 in perspective, recall from Section 6 and Table 6.1 that in-plane maneuvers require relatively small values of ΔV. For example, a ΔV of 0.3 km/s could allow the SMV to move from a 400 km to 1,000 km altitude orbit in the same orbital plane, so that it could release the satellites in orbits with different altitudes. Or, by changing altitude and then returning to the original orbit, the SMV could release the satellites at different places on the same orbit (see Section 6). By using a ΔV of 0.1 to 0.2 km/s, the SMV could maneuver in about 24 hours to release a second satellite halfway around the same orbit from the previous satellite.

However, maneuvers that change the orbital plane require considerably more propellant: a ΔV of 1.0 km/s corresponds to changing the inclination of an orbit at an altitude of 500 km by only 7.5°, and a ΔV of 2 km/s to 15°, which are relatively small out-of-plane changes.

A similar analysis would apply if the SMV were intended to carry propellant to refuel several satellites, assuming that a significant amount of propellant was to be delivered to each satellite. The numbers listed in Table 9.1 would also apply to the case in which the MSV was to deliver 300 kg of propellant to each of three satellites.

Inspector Satellites

Inspector satellites are satellites that approach other satellites and inspect them, by taking images or other data. They may be useful for maintenance of the satellite or for space law verification. Having an inspector satellite rendezvous with one satellite and then maneuver to rendezvous with a second satellite means that it must move between these two orbits. In general, this will require maneuvering the inspector satellite to change its orbital plane, to change the size and shape of its orbit, and to change its position with respect to the other satellite in the orbit.

Not surprisingly, designing an inspector satellite with enough maneuverability to inspect satellites in different orbital planes can lead to requirements for large increases in the mass of the inspector satellite.

Whether this increase in propellant mass is a problem depends on the mass of the empty inspector satellite (i.e., without propellant). If the satellite is small enough, the total mass of satellite and propellant may not be prohibitively large. For example, simple maneuvering microsatellites carrying sensors are being developed with masses of tens of kilograms. Even if the propellant mass resulted in a total mass several times larger, a small launcher could still launch the satellite and its propellant.

If the inspector satellite limited its activities to inspecting satellites in or near a single plane, the propellant requirements could be moderate. An example would be an inspector intended for geostationary satellites, since these all lie in or near the equatorial plane. A second example would be an inspector intended for a single plane of a constellation that consisted of multiple planes with multiple satellites in each plane (as would be the case for space-based missile defense interceptors and ground-attack weapons, and communication satellites in low earth orbit). Keep in mind that satellites with the same inclination may lie in different orbital planes that are rotated around the Earth's axis with respect to one another; this is important to take into account when determining the amount of propellant mass needed to move between one satellite and another.

To see the implications of out-of-plane maneuvers on the propellant mass required for an inspector satellite, consider an inspector satellite (with propellant for moderate in-plane maneuvering) with a mass m_s. Adding enough propellant to allow it to inspect two satellites in low earth orbits that lie in orbital planes separated by only 30° in inclination would require a propellant mass of 3ms (assuming conventional thrusters). Launching this satellite would require placing twice as much mass in space as launching two inspector satellites on two launchers into the two orbital planes.

As illustrated above, the propellant requirements increase rapidly for larger plane changes. Rendezvousing with two satellites in orbital planes separated by 90° would require a propellant mass of 39ms (e.g., the total launch mass would be 400 kg for an inspector satellite of mass 10 kg)—and would require placing 20 times as much mass in space as launching two satellites separately.

As noted above, if the mass of the inspector satellite is small enough, the overall launch mass of the satellite and propellant may not be prohibitively large. However, if an on-demand launch capability exists, such as an airlaunch capability, it may be more efficient to place an inspector satellite in the proper orbit once a particular need arose, rather than attempting to station a highly maneuverable inspector satellite in orbit.

The propellant mass could be reduced by using an ion thruster rather than a conventional chemical thruster, but as shown in Section 7, the maneuvers would take much longer. For missions that are not time critical this may be acceptable.

Section - 10

Elements of a Satellite System

Some of the intrinsic attributes of satellites make them vulnerable in ways that ground-based systems are not. Satellites in orbit move at high speeds (see Section 4), rendering collisions with even small objects disastrous. Satellites are nearly impossible to hide: just as satellites can view large swaths of the Earth, they are also visible to observers over large swaths of the Earth (see Figure 5.4). Moreover, once in orbit, a satellite's motion is predictable and it takes significant effort to appreciably change the orbit (see Section 6). Even small evasive maneuvers to escape an anti-satellite attack could add up to a prohibitively large effort, since an adversary can take multiple shots at the satellite. Satellites are also difficult to protect: Launch mass is at a premium, so armor and defensive measures come at some price. Some satellites, such as communications satellites, are designed to be easily accessed by users across the globe, a sensitivity that can be exploited to harm them or interfere with their operation. And essentially no satellite can now be repaired once damaged.

Satellite systems have a number of components, some of which make better targets than others. A satellite system comprises the satellite itself, the ground stations used to operate and control them, and the links between them. This section describes the components and their functions and how vulnerable and critical they are. We place particular emphasis on those elements that might be targeted and note that successful interference with a satellite system may not involve an attack on the satellite itself.

Satellites vary greatly in size. For example, commercial communications satellites can be large. The body of a Boeing 702 communications satellite, which was first launched in 1999, is seven meters long, and its solar panels extend to a length of 48 meters. The average Boeing 702 weighs nearly 3 tons when launched (this mass includes its stationkeeping propellant).[1]

Satellites can be small, as well. The SNAP "nano" satellite, constructed by Surrey Satellite Technology Ltd., is only 0.33 meters long, with a total mass of 6 to 12 kg, which includes a payload of up to 4 kg. This small satellite was placed in orbit in June 2000 and was able to maneuver, image, correctly keep attitude, and communicate with the ground.[2]

Satellite Components

All satellites have some basic elements, as outlined below and shown schematically in Figure 10.1.

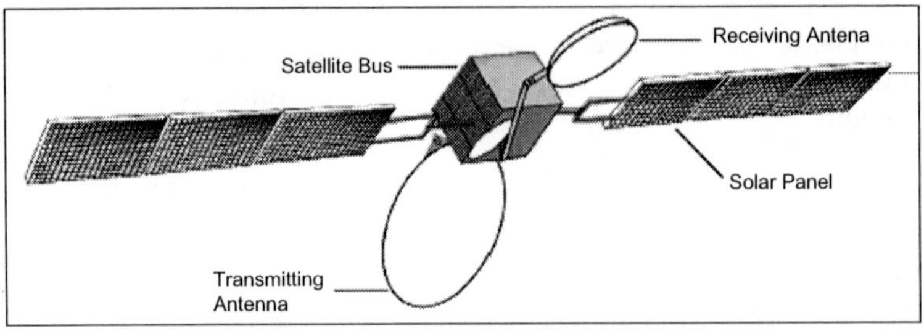

Figure 10.1. A schematic illustration of a satellite, including the satellite bus, the receiving and transmitting antennae, and the solar panels.

A structural subsystem, or bus. The bus is a metal or composite frame on which the other elements are mounted. Because it bears the stresses of launch, the bus is generally resilient. It may be painted with reflective paint to limit the solar heat it absorbs, which could also provide some protection from laser attacks.

A thermal regulation subsystem. This system keeps the active parts of the satellite cool enough to work properly. Active satellite

1. Boeing, "What Is a Satellite?" factsheet, http://www.boeing.com/defense-space/space/bss/ what_is_a_satellite.pdf, accessed December 15, 2004.

2. For information on SNAP, see http://zenit.sstl.co.uk/index.php?loc=47, accessed February 3, 2005.

components such as the computer and receiver can generate a large amount of heat. Sunlight incident on the satellite's surface also generates heat, although the satellite's surface can be made highly reflective to minimize heat absorption.[3] Without an atmosphere, conduction and convection cannot remove heat from an object as they do on Earth, so the satellite must radiate the heat to eliminate it. In most cases, the thermal regulation system is passive: just a set of well-designed thermally conducting pathways (heat pipes) and radiators to radiate the heat away. However, some components, such as some infrared sensors, may need cryogenic cooling; in this case, loss of the coolant would dramatically degrade the system's performance.

A large amount of heat introduced by an incident laser beam may be unmanageable: the internal electronics may fail if the bus conducts too much heat to them, or the structural integrity of the bus itself may be compromised.

A power source. Power is often supplied by arrays of solar cells ("solar panels") that generate electricity, which is stored in rechargeable batteries to ensure a power supply while the satellite is in shadow.[4] Technological improvements in battery technology have led to new battery types with high specific energy (energy stored per unit mass) and high reliability.

Solar cells are mounted on the body of a satellite or on flat panels. Mounting the solar cells on the satellite's body results in a more compact configuration (which may be desirable if space and mass are limited, or the satellite is meant to be covert), but since not all cells will be illuminated by the Sun at any one time, the power generated is less than it would be from large panels made of solar cells that are continually positioned to face the Sun.[5]

3. The incident sunlight that is not reflected is absorbed. If the satellite is disguised by limiting its reflectivity (i.e., painting it black), then it has a higher heat load.

4. For geostationary satellites, eclipses occur on 90 days a year, and last as long as 70 minutes (Bruno Pattan, Satellite Systems: Principles and Technologies [New York: Van Nostrand Reinhold, 1993], 26-29).

5. Solar panels that are properly oriented toward the Sun can provide about 130 W/m2 and 50 W/kg of power. Because solar cells mounted on the satellite's body will not, in general, be optimally oriented, they can typically provide 30 to 35 W/m2 and 8 to 12 W/kg of power (Gérard Maral and Michel Bousquet, Satellite Communications Systems, Fourth Edition [West Sussex, England: Wiley, 2002], 598).

The solar panels often have a large surface area compared with the rest of the satellite, so they sustain a relatively large number of collisions with debris particles. Solar panels are fragile and can be damaged easily, but partial damage to a solar panel may not disable the satellite:[6] satellites often can continue to function with partially working solar panels, albeit with diminished capacity. However, if the solar panels fail to deploy or are torn off, a satellite without another power source would cease functioning fairly quickly. A malfunction of the power distribution system could also totally impair the satellite.[7]

Other sources of power are available. The Soviet Union reportedly used nuclear reactors to power energy-intensive missions such as orbiting radar systems, and the United States launched one reactor-powered satellite.[8] Currently, the United States is considering a project to develop a uraniumfueled nuclear reactor to produce much higher levels of electric power in space.[9] On-board chemical sources of power are currently not used for satellites, although newer satellite designs may use fuel cells, which produce electricity by combining chemicals such as hydrogen and oxygen. Generators that produce electricity from

---

6. The Telstar 14/Estrela do Sul communications satellite failed to fully deploy one of its solar panels. Loral Space & Communications reports that the satellite generates enough power to maintain satellite health and to operate 17 of its 41 Ku-band transponders ("Loral To Initiate Limited Service On Telstar 14/ Estrela Do Sul In March," Loral press release, January 21, 2004, http://www.loralskynet.com/news_012104.asp, accessed December 15, 2004). On March 26, 1996, a solar panel on Canada's Anik-E1 satellite was disconnected, causing a power shortage and safety shutdown of the satellite. The satellite was restarted and was able to transmit a reduced number of television programs (Martyn Williams, "Galaxy IV Failure Highlights Reliance on Satellites," Government Computer News, May 20, 1998).

7. In September 2003, Loral Space & Communications declared the Telstar 4 satellite a total loss after it experienced a short circuit on the primary power bus ("Loral Skynet Declares Telstar 4 A Total Loss," Loral press release, September 22, 2003, http://www.loral.com/inthe news/030922.html, December 17, 2004).

8. Regina Hagen, "Nuclear Powered Space Missions—Past and Future," taken from Martin B. Kalinowski, ed., "Energy Supply for Deep Space Mission," IANUS 5/1998 Working Paper, http://www.globenet.free-online.co.uk/ianus/npsmfp.htm, December 15, 2004.

9. Current discussions call for a reactor capable of producing 100 kW of electricity (Ben Iannotta, "Jupiter Moon Probe Goes Nuclear," Aerospace America, March 2004, http://www. aiaa.org/aerospace/Article.cfm?issuetocid=474&ArchiveIssueID=50, December 15, 2004.

the heat released by radioactive materials (RTGs[10]) are currently used on deep space probes that move too far from the Sun to rely on solar panels. RTGs have been used on earth-orbiting satellites in the past but are not normally used on these orbits.

A computer control system. The on-board computer monitors the state of the satellite subsystems, controls its actions, and processes data. High-value satellites may incorporate sophisticated anti-jamming hardware that is operated by the computer. If someone gained control of the satellite's computer, the satellite could be made useless to its owners. Computer systems are also sensitive to their electromagnetic environment and may shut down or reboot during solar storms or if barraged by high levels of electromagnetic radiation.

A communications system. Communications form the link between the satellite and its ground stations or other satellites. This system generally consists of a receiver, transmitter, and one or more radio antennae.

The radio links between a satellite and the ground are one of the most critical and most vulnerable parts of a satellite system. All satellites require a link to and from the ground to perform "telemetry, tracking, and command" (TT&C) functions.[11] The TT&C system operates the satellite and evaluates the health of the satellite's other systems; it is therefore essential. The receivers on the satellite and on the ground can be overwhelmed by an intruding signal— called jamming—or confused by false signals—called spoofing. Although interfering with the TT&C channel could cause a great deal of damage, these channels are usually well protected with encryption and encoding. Generally, the more vulnerable piece of the communications system is that used for mission- specific communications, as discussed below.

10. Radioisotope Thermoelectric Generators (RTGs), which use the heat of radioactive decay of plutonium-238 to produce electricity, can produce up to hundreds of watts (Department of Energy Office of Space and Defense Power Systems, "Radioisotope Power Systems," http:// www.ne.doe.gov/space/space-desc.html, accessed December 15, 2004.

11. Telemetry refers to the information the satellite sends the control station about the status of its various components and how they are operating. Tracking refers to knowing where the satellite is; for example, the time for a signal to travel between the satellite and ground can be used to accurately determine the distance to the satellite. Command refers to the signals that are used to tell the satellite what to do.

The TT&C system occupies only a small part of the satellite's total assigned bandwidth.[12] A jamming attack would need to be mounted from the broadcast and reception area of the TT&C communications channel, i.e., the region from which a user can communicate with the satellite. Restricting the size of this area by increasing the antenna's directionality can help protect these channels from attack by reducing the region from which a jamming attack could take place. However, this may not be a viable solution for satellites that need to support users from a broad geographic area. Moreover, at a given frequency, improved directionality requires a bigger antenna.

An attitude control system. This system, which keeps the satellite pointed in the correct direction, may include gyroscopes, accelerometers, and visual guidance systems. Precise control is required to keep antennas pointed in the right direction for communication, and sensors pointed in the right direction for collecting data. If the attitude control system were not functioning, the satellite is unlikely to be usable.[13]

A propulsion subsystem. The satellite's propulsion system may include the engine that guides the spacecraft to its proper place in orbit once it has been launched, small thrusters used for stationkeeping and attitude control, and possibly larger thrusters for other types of maneuvering.

If the propulsion system does not function, because of damage or lack of propellant, the satellite may still be functional. However, in orbits dense with other satellites, such as geostationary orbit, satellites must be able to maintain their position very accurately or they will be a danger to their neighbors and to themselves. Satellites in low-altitude orbits need to make regular stationkeeping adjustments, without which their orbits will decay.

---

12. Bandwidth is the width of the band of frequencies that the satellite is assigned to use—the difference between the highest and the lowest frequency. The amount of data that can be sent through a band is proportional to the bandwidth.

13. In October 1997, trading on Bombay's National Stock Exchange in India was halted for four days after the Insat-2D satellite lost attitude control and began spinning in space. The problem was blamed on a power failure and cost the exchange around US$2 billion in losses (Martyn Williams, "Galaxy IV Failure Highlights Reliance on Satellites," Government Computer News, May 20, 1998).

Mission-specific equipment. In addition to the basic elements required for a satellite to operate, satellites also carry mission-specific equipment to carry out specific tasks. These may include

- *Radio receivers, transmitters, and transponders:* In addition to the communication equipment needed to operate the satellite, a satellite may carry similar equipment for other tasks. It may carry a radio antenna to collect radio signals, such as telephone or television signals, and to relay or rebroadcast them. The antenna serves to receive and transmit signals. It may be a parabolic dish (similar to satellite TV dishes), a feedhorn (a conical or cowbellshaped structure), or a minimal metal construction (similar to a rooftop TV antennae). When a system is designed to automatically receive a transmission, amplify it, and send it back to Earth, possibly at a different frequency, it is called a transponder.

  A satellite-based radar system is also composed in part of transmitters and receivers used to send and then receive the radio waves. Receivers are also used by the military for signals intelligence, i.e., eavesdropping on military communications, detecting the operating frequencies of enemy radar, or collecting telemetry from ballistic missile tests. Similarly, a satellite may carry transmitters to send out radio signals, such as the navigation signals from the Global Positioning System. A satellite may be designed to transmit a signal to a specific receiver on the Earth, or to broadcast it over a large area.

- *Remote-sensing systems:* The satellite's mission may be to take detailed images of the Earth's surface or atmosphere or objects in space, or to collect other types of data about the Earth and the atmosphere. A satellite may therefore carry such devices as optical cameras, infrared sensors, spectrographs, and charge-coupled devices (CCDs). For civilian scientific missions, these payloads are often complex, unique, and the result of many years of development.

- *Weapons systems:* A satellite may carry equipment to be used for attacking other satellites or targets on the ground or in the atmosphere. For example, it could carry a laser system and the fuel and mirrors needed to use it, or an explosive charge intended to destroy another satellite.

Ground Stations

Satellites are monitored and controlled from their ground stations. One type of ground station is the control station, which monitors the health and status of the satellite, sends it commands of various kinds, and receives data sent by the satellite. The antenna that the control station uses to communicate with the satellite may be located with the station, but it need not be: to maintain constant contact with a satellite not in geostationary orbit, and which therefore moves relative to the Earth, the station needs to have antennae or autonomous stations in more than one location.

Satellites may also have other types of ground stations. For example, a communication satellite's mission is to send data (voice communication, credit card authorization, video broadcast, etc.) from one user to another, and each user needs an antenna and is in effect a ground station. A satellite may therefore be communicating with many ground stations at the same time. For example, a Boeing 702 communication satellite can carry over 100 transponders.[14] Military communications satellites have ground stations that range from large, permanent command headquarters to small, mobile field terminals.

Ground stations are generally not highly protected from physical attack. Disabling a control station may have an immediate disruptive effect, but the disruption can be reduced by having redundant capabilities, such as alternate control centers. Computers at control centers may be vulnerable to attack and interference, especially if they are connected to the Internet. However, highvalue command computers will have high security, and many of the military command center computers are isolated from the Internet.

Links

The term link refers to a path used to communicate with the satellite (and is sometimes used to refer to the communication itself):

14. "Boeing 702 Fleet," http://www.boeing.com/defense-space/space/bss/factsheets/702/ 702fleet.html, December 15, 2004.

- Uplinks transmit signals from a ground station to the satellite.
- Downlinks transmit signals from the satellite to a ground station.
- Crosslinks transmit signals from satellite to satellite.
- Telemetry, tracking, and command (TT&C) link is the part of the uplink and downlink used to control a satellite's function and monitor its health.

The uplinks and downlinks are vulnerable to interference since the strength of the radio signals when they reach the receiving antenna is often low, so that an interfering signal need not be strong. Links can also be interfered with by placing something impermeable to radio waves, such as a sheet of conducting material, in the path between the satellite and ground station. This would likely be done close to the receiver or transmitter, where it could achieve the greatest effect.

Section - 11

Overview of Interfering with Satellite Systems

This section gives an overview of some of the means of interfering with satellite systems—both military and civil. Military satellites may be the most obvious targets, but civil satellites perform many essential support functions for military and political operations, such as communications and reconnaissance, and loss of some civil satellites could cause economic distress or enough disruption to make a political point.[1] Military and civil satellites may have different vulnerabilities to some kinds of interference, so that assessment of the tradeoffs associated with protecting them may differ markedly.

Anti-satellite (ASAT) attacks can take a variety of forms and serve a range of goals. For example, they may cause temporary, reversible interference, or they may be intended to cause permanent damage. They may target the satellite, the ground station, or the links between them. They may be overt, or they may be intended to be covert and thus not attributable to the attacker.

1. An estimated 60% of the military's satellite communications during U.S. operations against Afghanistan in 2001 (Operation Enduring Freedom) went through commercial satellites (Futron Corporation, U.S. Government Market Opportunity for Commercial Satellite Operators: For Today or Here to Stay? April 29, 2003, 4, http://www. futron. com/pdf/governmentwhitepaper. pdf, accessed December 17, 2004). The U.S. government was concerned about the availability of high-resolution commercial satellite images during the 2002 war in Afghanistan and used "shutter control"—buying exclusive rights to the Earth images taken in certain parts of the Ikonos satellite's orbits—during the first months of the conflict (John J. Lumpkin, "US Loses Edge on Spy Satellites," Associated Press, April 9, 2002). And state-sponsored satellite broadcasts in Iran and China have been jammed by political opponents. (Safa Haeri, "Cuba Blows the Whistle on Iranian Jamming," Asia Times, August 22, 2003; "China to Launch 'antijamming' Sinosat-2 satellite in 2005," Xinhua News Agency, March 4, 2004.)

The ASAT system may be based on the ground or in space. It may be relatively simple or require sophisticated technology appropriate to a space-faring nation. It may be able to interfere only with satellites in low earth orbit, or it may reach all the way to geostationary altitude.

Different methods of attack provide the attacker different levels of confidence of success. For some, success may be evident, while for others ascertaining whether the attack met its goal may be difficult. Some types of attack are easier to prevent or defend against than others.

This section includes information about these aspects of interference, organized (approximately) by the persistence of the effects. This arrangement traces fairly well the gradation from technically simple to technically demanding.

Preventing a satellite from accomplishing its mission temporarily, reversibly, or nondestructively is commonly called denial, while permanent disabling is called destruction. However, the distinction is not perfectly clear: whether a technique accomplishes denial or destruction can depend on a situation's details. For example, some denial techniques, such as dazzling a sensor with a laser and the use of high power microwaves to disrupt electronics, become destructive at higher powers; below we discuss reversible and permanent effects together for these systems.

Temporary and reversible interference with a satellite system is likely to be less provocative than destructive attacks. Such interference can, in some cases, be plausibly deniable. And it would not damage the space environment by generating debris. These techniques seem to be favored by military planners in the United States and elsewhere. Moreover, temporary interference with a satellite's mission, particularly over one's own territory, is likely to be perceived as defensive and legitimate in a way that permanently disabling the satellite would not.

Note, however, that impairing an individual satellite does not necessarily impair the mission of the constellation of which the satellite is a part. In a system that includes redundancy, back-ups, and alternatives in its design, the vulnerability of individual components need not lead to vulnerability of the system.[2] This point is discussed further in Section 12. The following discussion considers active interference with a satellite system, but some satellite missions can be frustrated with passive measures. For example, hiding, camouflaging, or moving valuable assets may deny a remote sensing satellite the ability to acquire information about them. Similarly, for satellites designed to attack ground targets or other satellites, adding protection to those objects can deny the satellite that ability.[3]

Electronic Interference : Jamming And Spoofing

As discussed in Section 10, satellites communicate with ground-based stations or receivers for a variety of purposes. Signals sent from the ground to the satellite are referred to as the uplink; those from the satellite to the ground as the downlink. Jamming refers to disrupting communication with a satellite by overpowering the signals being sent to or from the satellite by using a signal at the same frequency and higher power. The jamming signal may simply be meaningless noise that drowns out the real signal at the receiver. Spoofing, however, mimics the characteristics of a true signal so that the user receives the fake (or spoofed) signal instead of the real one.[4]

To be effective, the jammer or spoofer must be within the broadcast/receive area (the area from which broadcast signals can be sent so that they can be received by the receiver) of the receiver it is trying to jam, and it must be able to direct its signal to the receiver.

---

2. This issue is also discussed in Bruce M. DeBlois et al., "Space Weapons: Crossing the U.S. Rubicon," International Security 29 (fall 2004): 1-34.

3. For a discussion of denial and protection, see DeBlois et al..

4. To successfully spoof a receiver, the power of the spoofed signal at the receiver must be nearly the same as that of the true signal. If it is stronger, then the receiver will be jammed rather than spoofed; if it is weaker, the receiver will ignore the spoofed signal in favor of the true signal. T.A. Spencer and R. A. Walker, "A Case Study of GPS Susceptibility to Multipath and Spoofing Interference," 2003, www.eese.bee.qut.edu.au/QUAV/Unrestricted/ Postgraduate/ GPS%20 Interference/Conference%20Papers/AIAC_03.pdf, accessed January 16, 2005.

Jamming or Spoofing the Downlink

By jamming the downlink, an attacker prevents a ground station (or groundbased receiver such as a television, radio, or GPS receiver) from receiving a usable signal from the satellite. In the case of spoofing, the receiver receives a usable but false signal. Some receivers are designed to receive signals from satellites located anywhere in the sky, because this precludes the need to track the satellites or to orient the receiver in a certain direction and allows the receiver to be less complex and expensive. Placing a jammer or spoofer in the broadcast/receive area of such a receiver could be accomplished relatively easily by, for example, placing it on a hill or on an airplane.

The size of the area a jammer or spoofer affects depends in part on the power of the jammer and the strength of the satellite signal. A ground-based jammer or spoofer has the significant advantage of being much closer to the ground-based receivers than the satellite is, so the diminution of the signal by distance is not as pronounced and the jammer or spoofer needs to transmit much less power than the satellite does. And because the jammer or spoofer does not need to be positioned close to the receiver to produce a modest signal, the attacker does not need to know with great accuracy the location of the receiver(s) it is seeking to jam or spoof.

Simple jammers are inexpensive to make or to buy. For example, GPS jammers on the commercial market can reportedly interfere with receivers 150–200 km away, and instructions are available on the Internet for building a homemade GPS jammer inexpensively.[5] Spoofing devices are much more technically complex, since they must be able to mimic in detail the true satellite signal. However, GPS simulators that could spoof GPS receivers can also be purchased. (Jamming and spoofing of GPS receivers is discussed further in Section 12.)

Downlink jamming can be countered in several ways. Any antijamming technology makes it more difficult for an attacker to predict how effective jamming will be.

---

5. A New Scientist article describes an Air Force team that built a jammer to work against an ultrahigh frequency satellite, with just an Internet connection and $7,500 worth of materials. It would be fairly simple to adapt the same technique to the GPS frequency. (Paul Marks, "Wanna Jam It?" New Scientist, April 22, 2000.)

In some cases it may be possible to increase the power of the satellite's broadcasted signal. The downlink signal can be encoded, thus allowing the receiver to distinguish the real signal from the interference by comparing the incoming signal to a template known only to the user. However, these fixes add complexity and cost to the satellite and receiver system.

The receivers on the ground can be designed to receive signals only from the direction of the transmitters they are to communicate with and to reject signals from other directions. However, such antijamming features can increase the cost and weight of the receivers and, particularly for handheld receivers such as GPS receivers, may in the end make the receivers less usable.[6]

Another method to counter jamming is to have the satellite concentrate its power in a small frequency band and the receiver filter out all other frequencies. If the jammer does not know what frequency the system using, it must spread its power over a much broader range of frequencies to make sure it covers the frequency that is actually being used; such broadband jamming can require much higher power. This antijamming technique is a simple version of the more complicated signal manipulations performed in antijamming systems. Such systems may jump between frequency bands using a pattern known only to the legitimate user, making it difficult for the jammer to discover the frequency band being used for transmission fast enough to jam it. However, by forcing the satellite system to use only a small frequency band to transmit information at any one time, jamming or the threat of jamming can significantly reduce the rate at which information can be transmitted to and from the satellite even if it cannot stop transmision altogether.

In principle, a downlink jammer could be placed in low earth orbit to jam transmissions from satellites in high orbit. Since such

---

6. Many of the troops in the field in the 2003 Iraq war carried their own commercial Global Positioning System receivers rather than those issued by the military, because the commercial receivers were significantly lighter and consumed battery charge at a slower rate, or because they were not issued military receivers (Entry for May 8, 2003, http://www.coldsteelinfantry. com/ iraqi%20freedom%202. htm, accessed January 17, 2005, a resource website for the soldiers and families of the 2nd Battalion of the 113th Mechanized Infantry Regiment, 42nd Infantry Division, United States Army; Joshua Davis, "If We Run Out of Batteries, This War Is Screwed," Wired, June 2003, http://www.wired.com/wired/ archive/11.06/ battlefield_pr.html, accessed January 17, 2005).

a jammer would be 50 to 100 times closer to the receiver than a satellite in geosynchronous or semisynchronous orbit, it could generate significantly larger signals at the receiver. However, since the jammer would move rapidly with respect to the Earth, such schemes are likely to be impractical since they would require a large number of orbiting jammers to keep one in the receive area of the ground user. Increasing the directionality of the receiver's antenna would increase the number of jammers required.

Finally, if a jammer can be located, it can be attacked directly—which is likely to be seen as a legitimate action during a military crisis. A stationary jammer, particularly one sending out a strong signal to, for example, jam receivers over a large area, will be relatively easy to locate and disable and is likely to cause only limited interruption of communication. During the 2003 Iraq war, for example, the GPS jammers used by the Iraqi forces were readily identified and destroyed by the U.S. forces. (Section 12 discusses an alternative approach to GPS jamming that may be more difficult to counter.)

To counter spoofing, the signal from the satellite can be encrypted— scrambled before it is sent and unscrambled after receipt. Because sophisticated techniques such as encoding and encryption add complexity and reduce the amount of data the satellite can handle, commercial satellite operators are unlikely to find a financial case for adopting such techniques unless the threat scenario changes significantly. For sensitive military and other missions that require dependable and secure links, the tradeoff may, in some cases, make sense.

Jamming or Spoofing the Uplink

The receivers on the satellite itself can be jammed as well, preventing them from receiving the uplink signal. Satellites use uplink receivers to receive command-and-control communications.[7] These links are normally well protected from jamming by encoding the signal and from spoofing by encrypting the signal. Nevertheless, a high-power jammer can defeat the protection provided by encoding by essentially creating too much noise to sort through.

---

7. Satellites may use cross-links to perform these functions, but this is uncommon.

Communications and broadcast satellites use uplink receivers to get signals from the ground they will subsequently retransmit. While military satellites may encode or encrypt these signals before retransmitting them, commercial satellites often receive and retransmit data with a minimum of processing. Such rebroadcast satellites essentially route information from one point on Earth to another; for this reason, they are sometimes referred to as bent pipes. It is relatively easy to jam such bent pipe receivers: a ground-based jammer for such communications and broadcast satellites is basically a higher power version of standard communications equipment. Even satellites in geosynchronous orbits can be jammed from the ground, as both the jamming signal and the true signal it is trying to overpower have to travel the same distance and so experience the same decrease in signal strength due to distance.

Commercial communications and broadcast satellites may be particularly vulnerable to uplink jamming and spoofing for another reason: they are designed to receive signals from users over broad ground areas, and thus there will be a large area from which it will be possible to jam or spoof the uplink.[8] (Many such satellites are in geosynchronous orbits, and the broadcast/ receive area may cover a large fraction of the Earth's hemisphere.) Thus, a signal originating in one country could be jammed using a jammer in another country. In contrast, an attacker trying to jam the downlink signal from the satellite and to overwhelm the ground-based receiver would need to be somewhere near the receiver.

Jamming or spoofing attacks on commercial satellites could be a particular concern during a crisis for those countries that use commercial satellites to carry some or all of their military communications, including the United States.

Jamming communications broadcast satellites is not a purely theoretical threat. In July 2003, transmissions from the United States being broadcast via the Telstar 12 satellite to Iran were reportedly jammed by Iranians in Cuba, who used a ground-based

8. Geoffrey Forden, "Appendix B: Anti-satellite Weapons," in Ensuring America's Space Security, Report of the FAS Panel on Weapons in Space, (Washington, DC: Federation of American Scientists, 2004) 75-81, http://www.fas.org/main/content.jsp?formAction= 297&contentId=311, accessed January 17, 2005.

jammer to jam the uplink signals from the United States to the satellite. The United States had recently begun broadcasting its Voice of America program in Farsi, and several private Iranian-American groups encouraging protests against the Iranian government had increased their broadcast programs. The jamming was discontinued after discussions among the interested parties.[9]

As a second example, the China Central Television broadcasts of the 2003 Shenzhou V manned spaceflight were reportedly jammed by the group Falun Gong, which is believed to have repeatedly used transmitters in Taiwan to jam broadcasts in the Chinese mainland by jamming the uplink to the satellite. The new generation of Chinese communications satellites (Sinosat) will reportedly be carrying antijamming equipment.[10]

Jamming uplinks to satellites other than communications and broadcast satellites in geostationary orbits is technically more demanding, since the attacker needs to locate and perhaps track the satellite. This would be the case for any communications networks based in low earth orbit (such as the Iridium system) and for any satellite not in geostationary orbit.

Without detailed knowledge of the satellite, the jammer or spoofer may not be able to promptly determine the success of the attack on a command link, as many satellites can perform autonomously for some time and the behavior of the satellite would not change suddenly. It would be easier to determine the effect on a communications or broadcast satellite, as the downlink could be monitored for changes.

The antijamming and antispoofing techniques discussed above for uplinks could also be used to defeat downlink jamming and spoofing. However, it is generally not feasible for commercial satellites to use a directional antenna, since they rely on being able to serve customers from widespread locations. Nor is it always practical for military satellites, which must often accommodate a large number of users with different uplink capabilities (from mobile field terminals to permanent command centers) whose

9. Haeri. Iran has also placed strong restrictions on the ownership of satellite broadcast receivers, and the government jams many foreign broadcasts locally by jamming the downlink from the satellite.

10. "China to Launch 'anti-jamming'," Xinhua News Agency.

locations may be unpredictable and widely dispersed. Moreover, it may be more difficult to locate an uplink than a downlink jammer since, in many cases, an uplink jammer could operate from anywhere within a large area.[11]

Satellite-Based Uplink Jammers. Any space-faring country could in principle place an uplink jammer on a small satellite close to the target satellite. Because the distance from the jammer to the receiver would be hundreds or thousands of times smaller than the distance from the ground station transmitter to the receiver, the space-based jammer would need tens of thousands of times less power in its signal to give equal signal strength at the satellite receiver.[12] If the jammer were able to orient itself so that its signal could be received by the satellite's antenna, it might be able to conduct effective broadband jamming with low power.

However, it may be difficult in practice to make effective space-based jammers. For the jammer to be in the broadcast/ receive area of the satellite's antenna, it would need to be in an orbit below the satellite. Since its speed in that orbit would be greater than that of the satellite, it would quickly cross and move out of the antenna's broadcast/receive area.[13] Keeping a jammer in position to jam the satellite would require essentially continual maneuvering, significantly complicating operations.

A simpler arrangement would be to place a space-based jammer in the same orbit as the satellite, trailing it by a small distance, since it could then maintain a constant distance from the satellite. However, the jammer would be beside and not below the satellite, so it would not be in the main broadcast/ receive area of the satellite's antenna. Satellite antennae do have some sensitivity to signals coming from directions other than in the main broadcast/

---

11. Locating the source of satellite interference is the business of at least one company, Transmitter Location Services, LLC, based in Chantilly, Virginia. The company's website is http://www.tls2000.com, accessed January 17, 2005.

12. The signal strength decreases as one over the square of the distance from its source.

13. A jammer in an orbit 1 km below a satellite whose antenna was designed to view the entire section of the Earth below it would cross the broadcast receive area in 2 to 3 hours, whether the satellite was in low earth orbit or geosynchronous orbit. The jammer would spend proportionately less time in a smaller broadcast receive area.

receive area; these directions are covered by the side lobes of the antenna. However, the antenna's sensitivity in these directions is many factors of ten less than its sensitivity to signals coming from in front of the antenna. Moreover, once in orbit, the satellite may be able to control the shape and location of the side lobes to suppress them in the direction of a co-orbital jammer. This would result in lower sensitivity to jamming signals entering through side lobes and could easily eliminate the advantage of placing the jammer in space.

Laser Attacks on Satellite Sensors

Directed energy weapons, such as lasers and microwave weapons, have a number of desirable features for an attacker. The beams reach their targets rapidly since they travel at the speed of light, and the delivered power can be tailored to produce temporary and reversible effects or permanent, debilitating damage. Directed energy weapons also have disadvantages relative to physical interceptors: they can only reach targets in their line of sight, unless relay mirrors are used, and simple shields of reflective, absorptive, or conductive material can be effective defenses.

Lasers are especially useful for directed energy attacks because they can emit a large amount of energy in a narrow beam and a narrow band of frequencies. In principle, these features allow the attacker to efficiently direct energy to the right spot on a satellite with the proper frequency to inflict damage; in practice, however, the frequencies that can be used are constrained by available technology and other considerations, such as the need to choose a frequency that penetrates the atmosphere in the case of a ground-based laser. Moreover, if the attack requires energy at a range of different frequencies, either multiple lasers that produce different frequencies, or a broadband source may be required, as discussed below.

Lasers can attempt to interfere with a satellite's sensors or to damage the satellite by depositing a large amount of energy. The latter requires much higher power than the former and is discussed later in the section.

Laser technology is mature, and a variety of laser materials and techniques have been developed with a range of power levels. Lasers fall into one of two general categories depending on whether they produce power continuously (continuous wave (CW) lasers) or in short, repeated bursts (pulsed lasers). The distinction between the two is important for ASAT effects. CW lasers deliver a continuous stream of energy. A simple tabletop CW laser can generate from tens to hundreds of watts; large commercial CW lasers can generate tens of kilowatts or higher.[14] The U.S. Army's Mid-Infrared Advanced Chemical Laser (MIRACL) is a CW laser described as being in the megawatt range.[15]

Pulsed lasers can generate very high power levels over small fractions of a second (referred to as peak power), while having modest average power levels (when averaged over seconds). The pulse length and total energy per pulse are also important parameters. The highest power commercial pulsed lasers[16] can deliver terawatts of peak power but only in very short pulses, giving an energy per pulse (average power times pulse length) of 20 J; such pulse energies are common in longer pulses.

As laser power increases, the lasers become larger and more complicated, since they require large power supplies, cooling, and, in some cases, exhaust systems. For example, the MIRACL is fueled by a chemical reaction similar to that used in rocket engines and requires the support of a large facility. The Air-Borne Laser being designed for missile defense with a goal of having power in the megawatt range will have a mass of about 100 tons.[17]

---

14. A 10 kW 1.315 μm chemical oxygen iodine laser (COIL) costs about $10 million (see, for example, http://www.tokyo.afosr.af.mil/coil.html, accessed January 14, 2005).

15. The MIRACL laser is located at the High Energy Laser Systems Test Facility (HELSTF) at White Sands Missile Range, New Mexico. Its beam has wavelengths in the 3.6 to 4.0 μm range. (See, for example, the MIRACL system's homepage at the HELSTF website, http:// helstf-www.wsmr.army.mil/miracl.htm, accessed December 20, 2004).

16. Coherent Inc. builds high power commercial lasers, including an Nd:Glass pulsed laser that produces 40 terawatts (TW) of peak power with 20-J pulses. See "Multi-Terawatt Systems," http://www.coherentinc.com/Lasers/index.cfm?fuseaction=show.page&ID=726, accessed January 11, 2005.

17. "Extra Weight Will Not Affect ABL Test, Director Says," Global Security Newswire, March 7, 2003, http://www.nti.org/d_newswire/issues/newswires/2003_3_7.html, accessed December 20, 2004. The Airborne Laser System is intended to be carried by a modified Boeing 747-400F freighter airplane. See Boeing Integrated Defense System's Airborne Laser System webpage at http://www.boeing.com/defense-space/military/abl/flash.html, accessed December 20, 2004.

A laser ASAT system also requires a tracking and pointing system. A movable mirror can be used both to direct the laser beam toward the satellite and to focus the beam.

ASAT laser systems can be based on the ground, at sea, in the air, or in space. Ground and air-based laser ASAT systems would operate at visible and infrared wavelengths—wavelengths that can propagate through the atmosphere.[18] Powerful lasers can be readily made at these wavelengths, and light at these wavelengths can be aimed and focused at long distances using moderately sized mirrors. The atmosphere, however, is not perfectly transparent even at these wavelengths, and water vapor and other aerosols, as well as clouds and rain, will reduce the intensity of the beam.

Moreover, the ability to focus the beam may not be limited by the size of the laser's focusing mirror, but by the atmosphere. Turbulence in the lower atmosphere can disturb the transmission of the laser light and spread it out into a larger spot. For laser mirrors larger than a few tens of centimeters in diameter, the atmosphere limits how well the laser light can be focused.

Technical approaches to reducing atmospheric effects exist, such as adaptive optics, in which the mirror surface is rapidly deformed to compensate for atmospheric effects. These technologies are becoming more widespread, but they increase the complexity and cost of the mirror system. Moreover, using adaptive optics for this mission is more demanding than for typical astronomical uses.[19]

Dazzling

Lasers are commonly mentioned as being useful for interfering with satellites that take images of objects on the ground. This section discusses the utility of lasers for temporarily interfering with the sensor a satellite uses for such imaging; such temporary interference is called dazzling. Just as a satellite's receiver can be swamped by a jamming signal, a satellite's optical sensor can

---

18. The relevant atmospheric transmission windows are from about 0.35-0.9 μm (includes visible light and part of the near-infrared), 0.95-1.1 μm (near-infrared), 1.2-1.3 μm, 1.55-1.75 μm, and 2.0-2.3 μm (short-wave infrared), 3.5-4.1 μm (medium-wave infrared), and 8.0-13.0 μm (longwave infrared).

19. DeBlois, 58.

be dazzled by swamping it with light that is brighter than what it is trying to image. Remote sensing satellites that take high-resolution images of the ground have important strategic and tactical importance and thus may be attractive targets for sensor interference.

To understand dazzling, it is useful to understand how an imaging satellite works. The size of the ground area that the satellite's imaging system can see is determined by the field of view of the satellite's telescope and the size of its sensor. This area is generally much smaller than the total area that would be visible from the satellite (see Figure 5.4). For a satellite taking high-resolution images, this region is only tens of kilometers across. An attack on the satellite's sensor must originate from within the field of view of the satellite's telescope, or else the laser light cannot reach the detector.

The satellite's telescope and optical system focuses an image of a section of the Earth in the telescope's field of view onto a plane called the focal plane. On the focal plane is a sensor (or detector), frequently a device made up of a very large number of small, light-sensitive elements called pixels. Each pixel generates an electrical signal proportional to the intensity of the light that falls on it, and that signal is sent to a computer.[20] Part of the image on the focal plane falls on the detector; other parts of the image may pass through the telescope, but not fall on the detector. The portion of the Earth corresponding to the section of the image on the detector defines the detector's field of view.

The detector may be a two-dimensional matrix of pixels or a long linear array of pixels. In the first case, the detector tracks and receives light from one rectangular patch of the ground to produce an image, then tracks and receives light from the next patch of ground (this is called a step-stare system). This type of sensor is used in the Hubble Space Telescope (though, of course, not pointed towards the Earth), allowing it to stare at a given region of space so that it can collect sufficient light from dim celestial objects.

20. This is the reverse of how a television or computer generates images on the screen. The screen is made up of an array of tiny dots, and the computer creates an image by controlling the brightness of each of these dots. The more dots there are in the screen, the higher the resolution of the image.

Earth imaging satellites typically use linear arrays of pixels. As the satellite moves over the Earth, these arrays record the image a line at a time as they sweep over a continuous swath of the Earth. These individual images are stored and then stitched together by a computer to construct a two-dimensional image. (This is similar to the way scanners commonly used with home computers work.) This method of imaging is called pushbroom detection.

Thus each pixel corresponds to some small area on the ground within the satellite's full field of view and records the intensity of the light coming from that small area. The resolution of the satellite's imaging system is determined in part by how small an area on the ground corresponds to a single pixel, since the sensor will not be able to record variations of light and dark over smaller areas. For a satellite with a ground resolution of 1 m, for example, each pixel corresponds to regions on the ground about 1 m across.

For example, Space Imaging's IKONOS satellite, which has a ground resolution of 1 m, views a swath below it that is only 11 km wide. Its linear detector contains 13,500 pixels.[21] The French SPOT imaging satellite, which has a ground resolution of 2.5 m to 5 m, views a swath 60 km wide directly below the satellite.[22]

Consider a laser based on the ground that is attempting to dazzle a satellite with 1-m resolution. The laser uses a mirror to steer the beam and focus it on the satellite; that mirror might have a diameter of a few tens of centimeters. Since the satellite typically views a ground area that is tens of kilometers across, when this ground area is imaged on the satellite sensor, the mirror appears as a tiny point of light within that full area. If the satellite's optical system could create a perfect image of the mirror (ignoring effects of the atmosphere), that image would fall only on one or a few pixels on the detector, since each pixel would correspond to a ground area of 1 m across. In this case, if the laser is bright enough, it will dazzle those few pixels.

21. Gordon Petrine, "Optical Imagery from Airborne and Spaceborne Platforms," GEOInformatics, January/February 2002, 28-35. 22. "SPOT Image," http://www.spotimage.fr, accessed December

22, 2004. At an altitude of 832 km, the total observable ground area, as discussed in Section 5, would be a circle on the ground with a radius of roughly 3,000 km.

However, in a real (imperfect) optical system, the light coming from the laser is spread out over a larger part of the detector by several mechanisms. First, the fact that the satellite's telescope has a finite diameter leads to diffraction of the light, which spreads some of the light into a pattern of rings around the image of the laser's mirror. Second, small imperfections in any optical system tend to spread some fraction of the light passing through it out around the image. These imperfections have many sources, including errors in shaping or aligning the optics, distortions due to temperature gradients, and dust in the system. Finally, there may be bright reflections or glints off surfaces (such as edges of the optical elements) within the optical system. Satellite designers work to minimize this stray light and, under normal conditions, it may not be a problem. But stray light from a high intensity laser can be important.

The analysis in Appendix A to Section 11 gives a rough estimate of the laser power required to dazzle a portion of the detector of a high-resolution imaging satellite, assuming a ground-based laser with a 0.15 m diameter mirror. Note that the precise numbers depend on the laser wavelength used, the size of the laser mirror and satellite optical system, etc. Several important points emerge from that analysis. The first point is that because lasers can be focused into an extremely narrow beam, even low power lasers can dazzle small sections of a satellite's detector. However, if the satellite has high resolution, this section may correspond to a small region on the ground. For the situation considered in Appendix A to Section 11, a laser with a power of a milliwatt (mW)—roughly equivalent to that of a laser pointer[23]—appears to be able to dazzle a section of the detector corresponding to an area on the ground that is about 10 m in radius around the location of the laser.

Second, assuming the satellite's optical system is designed to control stray light as discussed in Appendix A to Section 11, the power required for dazzling larger areas of the detector increases rapidly. The calculation in the appendix suggests that laser power

23. Because the beam from a laser pointer has such a small diameter compared with the mirrors considered here for a laser ASAT, the beam would be much less intense when it reached the altitude of the satellite than the ASAT beams, and the laser alone could not be used for dazzling.

must be increased by a factor of 100 in order to increase by a factor of 10 the radius of the ground area obscured by dazzling. However, the required powers are within the range of commercially available lasers. The rough estimates in the appendix suggest that a 10 W laser could dazzle a region corresponding to a ground radius of about 1 km, and a kilowatt-class laser could dazzle a region with a radius of roughly 10 km. For a high-resolution satellite such as IKONOS, dazzling a 10-km region would dazzle essentially the full detector array.[24]

Third, the actual power levels required for dazzling depend on the details of the satellite's optical system. It may be possible to design an optical system with lower levels of stray light that would increase the power required to dazzle a given area. On the other hand, satellites not designed to deal with high light intensities might have much higher levels of stray light and could be much more sensitive to dazzling.[25]

The ability to generate these laser powers does not necessarily mean that an attacker can keep a satellite from viewing objects on the ground. Imaging satellites typically carry multiple detectors and filters. Each filter allows only a small band of wavelengths to pass through and reach one of the detectors. The multiple images of the scene taken at these different wavelengths can then be combined to give a full-color image of the scene (a table-top scanner produces color images in the same way). For example, the IKONOS satellite collects light in four bands. The discussion so far has assumed that the laser is operating at a wavelength that can pass through one of the filters and reach one of the detectors. If so, it will be able to dazzle that detector as discussed above. However, the filters greatly reduce the amount of light from that laser that can reach the other detectors. This attenuation of the beam greatly reduces the detector area that can be dazzled, or may eliminate dazzling on these other detectors altogether. Dazzling

24. At the large power required for dazzling a large area, the very intense light falling on pixels near the center of the diffraction pattern can damage those pixels, as discussed below.

25. During a test in 1997, a 30-W ground-based tracking laser reportedly dazzled an imaging satellite at 500 km altitude, although few details are available (John Donnelly, "Laser of 30 Watts Blinded Satellite 300 Miles High," Defense Week, December 8, 1997, 1).

large sections of all the detectors therefore requires the attacker to know the frequency bands of the various filters and to have lasers operating within each of these wavelength bands. If the attacker does not dazzle all the satellite's detectors, the satellite can still collect images of the ground.

Attempting to dazzle an imaging satellite with a space-based light source is difficult because of the requirement that the dazzler remain in the sensor's field of view, which is very small for the case of high-resolution imaging satellites. We do not consider this case in detail here.

To counter a dazzling attack, the satellite could change the direction it was looking or close a shutter to keep light from reaching the sensor. However, these both have the same effect as the dazzling attack: the satellite is unable to view the area of interest.

Partial Blinding

At sufficiently high intensities, laser light can permanently damage the sensors of imaging satellites. This report refers to such damage as partial blinding, since such an attack will damage only a portion of the sensor. The high intensity can cause the detector material to ablate or evaporate from parts of the detector. It can melt the material or its fragile electronic connections. In addition, the large temperature gradients produced by heat from the laser beam can produce thermo-mechanical stresses.[26]

Like dazzling, a blinding attack would need to be mounted from within the sensor's field of view for the laser to reach the satellite's detector. However, unlike dazzling, the laser needs to be within the field of view for only a very short time to damage the sensor.

The satellite's optical system helps concentrate the laser energy reaching the satellite by focusing it onto the detector. As with dazzling, this leads to very high intensity at the detector, but

26. A large body of literature investigating laser damage to detectors includes Vaidya Nathan, "Laser Damage in MWIR MCT Detectors," Proceedings of SPIE 2114 (1994): 726; F. Bartoli et al., "Irreversible Laser Damage in IR Detector Materials," Applied Optics 16 (November 1977): 2934–2937; Madhu Acharekar, et al. "Calculated and Measured Laser Induced Damage Threshold (LIDT) in Glass and Metal Optics," Proceedings of SPIE 3902 (2000): 85–96; and the literature referenced in Laser ASAT Test Verification (Washington, DC: Federation of American Scientists, 1991).

restricts the region of high intensity to a small part of the detector, so that it may damage only a few pixels. Unlike dazzling, however, the damage is permanent and cumulative: additional parts of the sensor may be damaged by subsequent attacks. For a linear pushbroom detector, damaged pixels will result in missing lines in the image as the detector sweeps over the swath of ground below it. Such attacks may strongly discourage the country owning the satellite from viewing the area where the laser is located.

The fact that the satellite's optical system concentrates the beam is important for estimating the laser power needed for blinding attacks. The concentration can be estimated by the ratio of the area of the satellite's telescope (which determines how much light is being collected) to the area on the detector onto which this light is focused. For high-resolution imaging satellites, this ratio can be greater than ten billion (see Appendix D to Section 11).

A linear detector in a high-resolution imaging satellite passing over a 1-m2 area on the ground or the mirror of a ground-based laser, collects light from that area on the ground for a very short time—a tenth of a millisecond— before moving on to adjacent areas. The laser would need to be powerful enough to deliver sufficient energy to damage the satellite's detector in that length of time. The required energy could be delivered either by pulsed lasers, which might have pulse lengths much shorter than a tenth of a millisecond, or by CW lasers, if they were powerful enough to deposit enough energy in the short time available.

Appendix D to Section 11 derives an estimate of the laser power required to damage a section of the detector of a high-resolution imaging satellite. This estimate is necessarily rough since it depends on details of the system, but it suggests a general scale of the power requirements. If the concentration of laser intensity by the satellite's optical system is high enough, even relatively low-power lasers—CW lasers with output powers of tens of watts or pulsed lasers with pulse energies of millijoules—appear to be capable of damaging small sections of a detector, corresponding to ground areas roughly 1 m in size (see Appendix D to Section 11).

Damaging a larger section of the detector requires considerably higher power. The estimates in the appendix suggest that increasing the size of the damaged region by a factor of 10 requires increasing the laser power by a factor of 100. These estimates suggest that power levels available from commercial lasers could damage sections of a detector corresponding to tens of meters on the ground.

Partial blinding by a laser based in space is possible since the laser needs to be in the satellite's field of view for only a very short time. The shorter distance between the laser and the satellite in this case would reduce the laser power required relative to a ground-based laser.

Shutters can in principle protect satellite sensors from blinding, although the satellite system would need to detect the attack with enough time to react. For example, the satellite could include a sensor with low sensitivity to survey the ground area ahead of the primary sensors[27] or might detect a lower-power aiming phase prior to the attack. A nonlinear optical material that becomes opaque to beams with high intensity might be placed at an intermediate focus in the optical path to act as a switch to protect the sensor.

As discussed in the dazzling section, a laser operating at a particular wavelength could only deliver high intensity laser light to a satellite's detector if that wavelength fell within the small band of wavelengths that could pass through one of the detector's optical filters; otherwise, the intensity of the light reaching the detector would be sharply reduced. To damage the detector, the attacker would therefore need to know the filter bands. Even if this was known, a laser operating at a single wavelength could damage only one of the multiple detectors an imaging satellite would carry.

Predicting and confirming the success of a blinding attack may not be simple. Blinding is more difficult to perform confidently than dazzling. The energy needed for a dazzling attack can be determined by the local conditions and the sensor's resolution on the ground: the dazzling laser just needs to be brighter than the light reflected by the Earth within that area. In contrast, the amount of energy required for blinding can vary by large factors depending on details of the satellite's optics and sensor system.

---

27. Ashton B. Carter, "Satellites and Anti-Satellites: The Limits of the Possible," International Security 10 (Spring 1986): 46-98.

For commercial and civil satellites, an attacker can gather details about the optical system and sensor design from public sources. For sensitive military and intelligence satellites, some information about the optics system (a modest guess as to the telescope aperture and focal length) can be gleaned from ground-based images of the satellites, and the sensor wavelengths can be surmised since the atmosphere limits the wavelengths that can be used and the wavebands are usually made as wide as possible so to increase the signalgathering capability. Sensor materials and technology are well understood and a determined adversary may be able to make reasonable estimates of the effects a laser would have on them, although the details of sensor construction could be important.

High-Power Ed Microwave Attacks

A second directed energy weapon that could be used to attack satellites is a device that produces high-powered microwaves (HPM). Microwaves are electromagnetic waves with wavelengths shorter than radio waves but considerably longer than visible light.[28] They are commonly used by radars and for sending communication signals.

HPM attacks could in principle be directed at a satellite either from a ground-based or space-based HPM weapon. Ground-based HPM weapons would have to contend with long distances to the satellite, which limits their utility. Producing high intensity at the satellite requires high levels of emitted power and a large antenna for focusing the beam.[29] Moreover, the atmosphere limits the transmission of beams of microwaves with very high power.[30] For

28. Microwaves are typically considered to lie in the frequency band between about 1 GHz (corresponding to a 30-cm wavelength) and 300 GHz (corresponding to a 1-mm wavelength).

29. Since microwaves have wavelengths thousands of times longer than optical light, focusing them is more difficult than focusing optical light since it requires a much larger antenna. This limits the ability of an HPM system to focus radiation over long distances. See Appendix B to Section 11 for a discussion of antenna size and directionality.

30. At high intensities, microwaves will cause the air to break down and no longer transmit. For lasers, which have shorter wavelengths, the atmosphere will transmit beams that are thousands of times more intense before breaking down (Philip E. Nielsen, Effects of Directed Energy Weapons, [Washington, DC: National Defense University, 1994], http://www.ndu.edu/ ctnsp/ directed_ energy.htm, accessed December 21, 2004).

these reasons ground-based HPM anti-satellite systems appear less interesting than HPM weapons that attack at shorter ranges: those based in space or popped up using a suborbital missile.

Microwave radiation at high intensities, if they are able to enter and affect (or couple to) some component of the satellite, can disrupt a satellite's electronics and, above some threshold, permanently damage them. In a nondestructive attack, the microwaves may, for example, reset computers and garble commands, disrupting the satellite's function during the attack and for a time after. HPM attacks can permanently damage a satellite's electronics if the strength of the microwaves that couple to the system is large enough.

The coupling of HPM to the satellite's electronics is characterized either as back door or front door. Front door attacks couple to the satellite through the antennae used for broadcast and communication, which are designed to receive and amplify radio signals with frequencies in or near this range. Front door attacks are therefore mounted from within the area in which the satellite can broadcast and receive signals. Unlike jamming, however, HPM attacks use a short, high-power pulse and need be in the broadcast/receive area only briefly.

Microwaves can couple through the front door at any frequency that the satellite's receiver system accepts; the success of coupling is therefore more predictable if this information is known. The receiving electronics in the satellite are often designed to pick up faint signals, and overwhelming them with high intensity radiation can leave them permanently damaged if they are not properly protected. While front door attacks can potentially couple a large amount of energy to the satellite, if the satellite is designed to detect and block large signals from reaching the sensitive components, delivering an effective attack may be difficult. The effect on a satellite will not be predictable without information about the satellite's design.[31]

31. Damage thresholds are very uncertain, as they depend on the details of the system being attacked. H. Keith Florig estimates (in "The Future Battlefield: A Blast of Gigawatts?" IEEE Spectrum 25 [March 1988]: 50-54) that a fluence (energy per area) in the front door of about 100 J/m2 would damage unshielded electronics that were directly coupled to the satellite's antenna. Other damage threshold estimates are given in Nielsen and the references therein.

In back door attacks, the microwaves enter the satellite by some other means than an antenna. The metal casing of a satellite helps shield its electronic components from microwave attacks, but microwaves can enter the satellite through small seams in the casing or gaps around electrical connections. If microwaves enter the satellite in this way, they can interact with and damage a wide variety of the electronics inside the satellite. Since back door attacks do not enter the satellite through the antenna, they need not take place from the broadcast/receive area of the satellite and they need not be in the frequency band the satellite is built to receive.

The amount of coupling and the effects of back door attacks are, however, difficult to predict and will be a major source of uncertainty to the attacker. The ability of microwaves to find ways to enter the satellite may depend on factors such as the quality of construction and effects of aging. The frequency of microwaves that can couple through the back door will not be known since it depends on the dimensions of these openings. Coupling to the target can be increased by chirping the microwaves—emitting over a range of frequencies— to increase the chances that one of the frequencies will couple to a back door. However, spreading the power over a range of frequencies decreases the power at any one frequency, which may also limit the weapon's effect. Since the microwave signal is not collected and amplified as it is in a front door attack, the power levels required for a successful back-door attack are significantly higher.[32]

HPM technology is still maturing. The principal technical issues are generating high power from modestly sized devices and packaging the emitter in a useable and robust platform.[33] One of the more developed and compact HPM sources (called a vircator) can reportedly generate tens of gigawatts of microwave power at frequencies up to the gigahertz frequencies used by satellite communications. One report states that a 400-kg device could

32. Florig estimates that a back door attack might require a fluence a thousand times higher than a front door attack to cause the same disruption.

33. See reviews by Carlo Kopp, "The Electromagnetic Bomb—a Weapon of Electrical Mass Destruction," Air & Space Power Chronicles, http://www.airpower.maxwell.af.mil/ airchronicles/ kopp/apjemp.html, accessed January 15, 2005, and Carlo Kopp, "An Introduction to the Technical and Operational Aspects of the Electromagnetic Bomb," Australian Air Power Studies Centre Paper 50, November 1996.

produce 2 to 5 gigawatts (GW) of HPM power in a short pulse.[34] This type of device generates its power using an explosive generator and so would be used only once. Such a weapon would have relatively short range: using a 1-m focusing antenna, it would need to be within about 1 km of an unshielded computer to disrupt it.[35] For a back door attack, poor coupling would decrease this distance; for a front-door attack, the distance could be tens of kilometers.

While the technology to create HPM exists and is likely to become more widely available, the effectiveness of these weapons will continue to be highly uncertain, which limits their utility, especially against targets considered important enough to attack. Both the extent to which the microwaves couple to the satellite components, and their effects on the components if they do couple, will be uncertain. A given attack could be destructive, disruptive, or completely ineffective, depending on the details of the satellite's design. Although classes of systems should respond similarly to an HPM attack, it is difficult to predict with any certainty how any particular satellite will react without actually testing it. Moreover, electronics can be hardened against microwave attacks of moderate levels without great cost if the protection is incorporated into the initial system design; a hardened satellite could withstand orders of magnitude higher HPM flux than an unhardened satellite.

Space-based HPM weapons would be available only to space-faring countries. Since the HPM weapon needs to be close to the satellite, a co-orbital weapon is likely to be detected and identified as a threat. However, an HPM weapon in a crossing orbit intended for use as it passed close to the target satellite might not be recognized as a threat. Countries with short-range missiles could attempt to loft an HPM weapon to a high altitude and set it off near a satellite it wanted to attack.

The attacker would need to be able to orient the weapon to aim it at the target satellite, and it would have to pass close enough to the satellite to be within the weapon's lethal range;[36] both of these factors could further increase the uncertainties in using the weapon.

---

34. Kopp, "An Introduction."

35. This range assumes 105 J/m2 is required to disrupt an unshielded computer and that the weapon creates a millisecond pulse (Florig).

36. For a discussion of the difficulties of placing a lofted payload near an orbiting satellite, see Section 12.

The attacker may not be able to immediately determine if the attack was successful, apart from monitoring downlinks. If the satellite was permanently disabled, this might become evident over the course of a few weeks if the satellite's stationkeeping maneuvers could be monitored.[37] If the satellite was not updating its orbit correctly, the attacker might surmise that it was no longer functioning.

Destruction

Attacks on Ground Stations

Satellite operators command satellites from ground stations, which may be attacked with weapons from the outside, by agents from the inside, or remotely by hackers. Precautions that a satellite owner could take include screening employees' backgrounds, physically protecting the station with walls and gated entry, and making plans to transfer the ground station's operations to another facility in an emergency.

A successful attack on a ground station is likely to be disruptive for a period of time, but with proper planning by the satellite's operator, use of the satellite should be restored relatively quickly, by, for example, transferring control of the satellite to a backup station. Unlike a damaged satellite, damage to the ground station can be repaired.

Laser Attacks on Satellites: Heating and Structural Damage

High-power lasers can subject satellites to large amounts of energy. The resulting heat can upset the delicate thermal balance of the satellite for long enough to damage the satellite's components or, if sufficiently intense, can damage a satellite's structure by, for example, weakening the hulls of pressurized tanks. Solar panels are also vulnerable to laser attacks.[38]

---

37. For an example of how this could be used to determine the operational status of a satellite, see Pavel Podvig, "History and the Current Status of the Russian Early Warning System," Science and Global Security 10 (2000): 21-60, http://www.princeton.edu/%7Eglobsec/ publications/ pdf/10_1Podvig.pdf, accessed January 10, 2005.

38. Damage of solar panels due to heating is discussed in FAS, "Laser ASAT Test Verification," 28 and in Forden, 75. Potential damage due to unequal charge across the panel is mentioned in Martin Unwin, "A Study into the Use of Laser Retroreflectors on a Small Satellite," 1995, http://www.ee.surrey.ac.uk/SSC/CSER/UOSAT/IJSSE/issue1/unwin/unwin.html, accessed January 15, 2005.

Since these attacks are not aimed at the sensor, the attacker is no longer confined to the satellite sensor's field of view, as in an attack intended to dazzle or blind. An attack can be mounted from any location that puts the satellite in the attacker's line of sight—from the ground, the air, or space—if sufficient laser power is available.

Compromising robust satellite components, such as the bus or non-sensor payload, requires powerful lasers. Studies of laser attacks on satellites estimate that for unshielded satellites in low earth orbits, ground-based megawatt class lasers could create this damage in a few seconds, and for the most fragile parts, kilowatt-class lasers could do the same in a longer period of time.[39] Laser attacks intended to disrupt the satellite by heating may require lower power. The altitude of geostationary satellites protects them from structural damage by lasers on the ground or in low earth orbits.

Developing a laser ASAT system for these kinds of attack is difficult and expensive, thus such attacks are restricted to technically sophisticated countries. Delivering high laser intensity to satellites requires a powerful laser, a large mirror for focusing the beam, and for ground-based lasers, adaptive optics to reduce atmospheric effects.[40] Currently, the technology does not exist to build a high-power space-based laser weapon.

There are some defensive measures a satellite could take, such as hardening exposed surfaces, building in redundancy, and deploying a protective shield against the laser light. (Satellites do not routinely carry shields today.) Such measures could allow the satellite to withstand the effects of the attack or could delay the onset of disruptive or lethal effects long enough to allow it to take other defensive actions. If the attacking laser were space-based, increasing the time the satellite is protected is particularly significant since requiring the laser to operate for longer periods of time could exhaust the supply of fuel the laser needs to create the laser beam.

---

39. Detailed technical analysis of using high-powered lasers against space targets can be found in "Report to the American Physical Society of the Study Group on Science and Technology of Directed Energy Weapons," Reviews of Modern Physics 59 (July 1987): S1-S201 and in FAS, "Laser ASAT Test Verification."

40. Large mirrors are more technically difficult to produce and much more expensive than smaller ones. The cost of a ground-based telescope rises with diameter at approximately the power of 2.5; see for example, Aden and Marjorie Meinel, "Extremely Large Sparse Aperture Telescopes," Optics and Photonics News, October 2003, 26-29.

The success of laser attacks intended to cause structural damage to an unshielded satellite should be fairly predictable; success on a satellite that has taken defensive precautions may be unpredictable. Similarly, the effectiveness of other kinds of laser attacks, such as attacks intended to disrupt the satellite by heating it, may be highly uncertain. As with other types of attacks, its effectiveness may be difficult to assess. Structural damage to a satellite in low earth orbit may be visible from the ground using a telescope. Otherwise, the attacker may need to rely on monitoring changes in the downlinks or in the satellite's stationkeeping maneuvers.

Kinetic Energy Attacks

Attacks that attempt to damage or destroy a satellite through high-speed collisions with another object are called kinetic energy attacks. Kinetic energy is the energy in the motion of an object. The faster two objects are moving relative to one another, the more kinetic energy is available to be turned into destructive force when they collide. Since satellites move at high speeds, a collision with even a small object can seriously damage them.[41] Even a collision that leaves the satellite largely intact could cause it to tumble.

Ground-based Kinetic Energy Attacks. Kinetic energy attacks that are launched from the Earth and attempt to destroy the satellite without placing an object into orbit are referred to as direct-ascent attacks.

Such an attack may use a homing interceptor. The ASAT would be launched on a missile that carries it above the atmosphere and releases it in the direction of the target satellite. The interceptor would then use its sensors to detect the target satellite and its thrusters to guide it to collide with the satellite. Shortly before intercept it might release a small cloud of pellets to increase the possibility of collision. Since the attack can be direct-ascent and does not require the interceptor to be placed in orbit, attacking satellites in low earth orbit requires only a relatively short-range missile to loft the interceptor to the satellite's altitude (see Section 8). Because of the difficulty of launching objects directly to

41. The kinetic energy of an object with mass m and speed V is given by mV2/2 and, therefore, increases rapidly with the speed of the object. The kinetic energy of an object of mass m traveling at 3 km/s is roughly the same as the explosive energy of the same mass m of high explosive.

geosynchronous altitudes, direct-ascent attacks are likely to be used only against satellites in low earth orbit.

For a homing ASAT, the attacker need not determine the trajectory of the satellite with high accuracy. It would need only to determine it accurately enough and deliver the interceptor into space accurately enough that the sensors on the interceptor could locate the satellite. The interceptor would also need to be close enough to the satellite so that its maneuvering capability is sufficient to guide it to intercept. The attacker would need the technical sophistication to build a simple homing device (which would include a sensor and the ability to maneuver accurately). If it can do these things, this method of attack has a high chance of success. Any space-faring country should be able to develop such an interceptor; simple homing systems using commercially available sensors should be within the reach of many other countries.

The United States successfully tested a direct-ascent interceptor against a satellite in low earth orbit in the 1980s. The interceptor was launched by a missile carried on an F-15 aircraft.[42] The interceptors being deployed as part of the U.S. ground-based midcourse missile defense system use direct-ascent kinetic energy interceptors to target ballistic missiles, and these could also be used to attack satellites throughout low earth orbit.[43]

If an attacker does not have the ability to develop a homing interceptor, but does have ballistic missile technology, it could instead try to launch a large cloud of pellets into the path of the satellite. The success of such an attack depends on various parameters, such as how well the attacker is able to determine the satellite's trajectory, how well the attacker can control the placement and dispersal of the pellet cloud, and what total mass of pellets the attacker's missile can loft to the orbital altitude of the satellite. This method is examined in more detail in Section 12, which finds that this method is unlikely to be an attacker's choice if other options exist. In particular, unless the attacker has

---

42. Laura Grego, "A History of US and Soviet ASAT Programs," April 9, 2003, http:// www. ucsusa.org/global_security/space_weapons/page.cfm?pageID=1151, accessed January 17, 2005.

43. See David Wright and Laura Grego, "Anti-Satellite Capabilities of Planned US Missile Defense Systems," December 9, 2002, http://www.ucsusa.org/global_ security/ space_ weapons/ page.cfm?pageID=1152, accessed January 17, 2005.

accurate missiles and good tracking capability, the effectiveness of such an attack may be low, and the attacker could have little confidence in the attack.

Some protection against attacks by small pellets can be gained by deploying shielding on the forward portion of the satellite, and modest maneuvering may be effective against a non-homing attack (see Section 12). However, neither shielding nor maneuvering capability could be expected to protect against homing interceptor attacks, since the mass of the objects hitting the satellite would likely be too large to shield against,[44] and the interceptor is likely to have more maneuvering capability than the satellite.

Damage from a kinetic energy attack to a satellite in low earth orbit is likely to be detectable from Earth using even a moderate-size telescope to image the satellite. Damage might also be assessed by monitoring the satellite's downlinks or tracking its stationkeeping maneuvers.

Destruction of a satellite by impact is likely to generate some persistent debris; just how much and how long the debris persists depends on the altitude of the satellite and the details of the collision. If the attacker has longterm interests in space, debris production may be a deterrent to using these types of weapons if other weapons are available.

Space-Based Kinetic Energy Attacks. Instead of using a direct-ascent approach, kinetic energy ASATs can also be placed in orbit prior to an attack. They may be launched shortly before the attack, as was the case with the co-orbital ASAT developed and tested by the Soviet Union in the 1960s to 1980s, which was intended to complete only a few orbits before attacking.[45] Space-based ASATs may also be deployed in space well before they are used; such ASATs are often called space mines (although the term includes ASATs that use other attack methods besides kinetic energy—see Section 12).

Like direct-ascent ASATs, space-based ASATs can use unguided clouds of pellets, or homing interceptors. A cloud of

44. A homing interceptor that released a cloud of pellets as a kill enhancer shortly before intercept could use fewer, larger pellets than in a nonhoming attack, since the cloud would be much smaller.

45. Grego.

pellets would be released in an orbit that crosses the satellite's orbit or in the same orbit as the target satellite, but moving in the opposite direction so that the relative speed in a collision would be large. An orbiting pellet cloud may be more effective over time at destroying a satellite than a lofted cloud, since the orbiting cloud could be designed to pass near the satellite repeatedly; however, since it would constitute persistent orbital debris, it could threaten other satellites as well over time.

A homing interceptor could be placed in a crossing orbit to allow a highspeed collision. Or it could be placed in the same orbit, in which case it could approach the target satellite slowly and, for example, fire a small cloud of pellets to destroy it.

An ASAT placed in a low-altitude orbit could also be used for a kinetic attack on satellites in much higher orbits, including geostationary orbits, if it is given sufficient propellant for maneuvering. In particular, the discussion in Section 6 shows that an ASAT in orbit at 400 km could reach a satellite in geostationary orbit if it is designed to have a ΔV of 2.4 km/s; the travel time to orbit in this case would be about 5 hours. If the ΔV of the ASAT is instead 4 km/s, the travel time to orbit would be reduced to 1.5 hours. This issue is of interest since the space-based interceptors (SBI) that might be placed in orbit as part of a space-based ballistic missile defense system would require a ΔV of this magnitude to allow them to swiftly engage a missile in its boost phase (see Section 9). The SBI could therefore be designed to have ASAT capability (if they are given the proper sensors, for example),[46] and a missile defense system would contain thousands of SBI—many more than the number of potential targets in high orbits. Even a relatively small number of SBI would represent a significant threat to satellites in geostationary orbit.[47]

46. Wright and Grego. kg.

47. Since an interceptor designed for attacking a satellite rather than a boosting missile would require less maneuverability for the homing process and for accelerating out of orbit, it could have significantly smaller mass. For example, using the same assumptions for the SBI as in the APS Boost-phase study (Report of the American Physical Society Study Group on Boost-Phase Intercept Systems for National Missile Defense, July 2003, http://www.aps.org/public_affairs/ popa/ reports/nmd03.html, accessed January 16, 2005) but with ΔV of 0.5 km/s for homing and 3 km/s for accelerating out of orbit (instead of 2.5 and 4 km/s), the SBI mass would decrease from 820 kg to 300

Because space-based ASATs must be placed in orbit, they are limited to countries with a space-launch capability. We discuss the possible advantages and disadvantages of ground- and space-basing for ASATs in more detail in Section 12.

Since both the target satellite and the kinetic energy ASATs considered in this section are in orbit, the speeds involved in a collision can be very high. While the speed and geometry of such collisions are likely to create persistent space debris, the amount and lifetime of that debris would depend on the details of the collision.

Bodyguard Satellites. Defensive satellites, often called bodyguard satellites, are sometimes discussed as a potential means of protecting high-value satellites from kinetic energy attacks.[48] How difficult this task is depends in part on what the bodyguard is intended to defend against. Defending against a co-orbital ASAT that approaches the satellite slowly might be relatively straightforward. Defending against ASATs in crossing orbits, however, could be difficult because the ASAT could approach at high speed and from a wide range of directions. A system of defensive satellites would require a capable surveillance system that could provide sufficient warning of an attack. For bodyguards using kinetic energy interceptors, even a successful intercept by the bodyguard could create debris that might damage the satellite it was defending.

Even if bodyguard satellites could be made to work against some kinds of threats, countries will not be able to rely on them to protect their satellites from direct attack or interference by a determined adversary. Their effectiveness against real-world attacks would not be known and they cannot be designed to defend against all possible threats to the satellite. Deploying bodyguard satellites does not preclude the need to take into account the vulnerability of satellite systems and to have back-up systems for any essential military capabilities provided by satellites.

---

48. See, for example, William L. Spacy II, "Does the United States Need Space-Based Weapons?" School of Advanced Airpower Studies, Air University, Maxwell Air Force Base, AL, 1998, 40–44, http://www.au.af.mil/au/awc/awcgate/saas/spacy_wl.pdf, accessed January 16, 2005; and DeBlois, 60.

Electromagnetic Pulse from a High-Altitude Nuclear Explosion

A nuclear explosion at an altitude of several hundred kilometers would create an intense electromagnetic pulse (EMP) that would likely destroy all unshielded satellites in low earth orbit that are in the line of sight of the explosion.

In addition, the explosion would generate a persistent radiation environment that would slowly damage unshielded satellites in LEO.[49] The radiation environment could also make it more difficult for high-altitude satellites to communicate with ground stations (depending on their communication frequencies) and would last months to years. However, the extent of the damage the increased radiation could cause is uncertain, and shielding satellites against it is estimated to add only a few percent to the cost of the satellite. The designs of many military satellites incorporate protections against EMP and increased radiation.

Such an attack is indiscriminate and unlikely to be undertaken by an adversary with investments or aspirations in low earth orbit. However, because of the large area it could affect and the persistence of the radiation effects, detonating a nuclear weapon in space could be a highly effective terrorist-style attack by a country that had a nuclear weapon and a medium-range missile to launch it.

In principle, a nuclear warhead intended as an ASAT could be launched from the ground or based in space. However, the Outer Space Treaty prohibits the signatories from placing nuclear weapons in orbit. Moreover, a country with only a few nuclear weapons seems unlikely to place one into orbit for possible future use against satellites.

---

49. For a discussion of these issues, see FAS, Ensuring America's Space Security, 23; and Defense Threat Reduction Agency, "High Altitude Nuclear Detonations (HAND) Against Low Earth Orbit Satellites (HALEOS)," briefing slides, April 2001, http://www.fas.org/ spp/ military/ program/ asat/haleos.pdf, accessed January 16, 2005.

Appendix - 11 A

Laser Power Required for Dazzling

To dazzle an earth-observing satellite, the dazzler needs to produce a signal at the sensor stronger than the light reflected from the Earth. The brightness of this reflected light varies widely by time of day and by the surface from which it reflects.[1]

The intensity of sunlight reaching the surface of the Earth in a 1-micrometer (μm) band around a wavelength of 1 μm is roughly[2] 600 W/m^2. To get a rough estimate of the intensity at the satellite of the sunlight that is reflected from the Earth, we assume that the full incident intensity is reflected diffusely, so that it is spread over 2 steradians (sr). Under these assumptions, the scattered sunlight from 1 m^2 of the Earth is roughly 100 W/sr. For this calculation, the satellite is assumed to have a ground resolution of 1 m and is assumed to image a 1-m2 piece of Earth onto one pixel.

Now consider a laser on the ground of power P watts, operating at a wavelength of 1 μm, and assume the laser beam is focused by a mirror with diameter D_L meters. The diffraction limit of the mirror allows the beam to be focused into a solid angle of approximately $(1.22\lambda/D_L)^2$ sr (see Appendix B to Section 11). Atmospheric effects also cause the beam to spread, and this effect may be larger than the mirror's diffraction. At optical wavelengths, the atmospheric effects are such that, unless adaptive optics are used, increasing the size of the mirror beyond about 0.15 m does not result in a more compact beam. A mirror of this size focuses the laser power into a solid angle of roughly 10–10 sr.

1. A sensor may have settings that make it is less responsive to light when it is viewing bright areas and more responsive when viewing dark areas.

2. S.C. Liew, "Principles of Remote Sensing," Tutorial, http://www.crisp.nus.edu.sg/~research/ tutorial/solrdn.gif, accessed January 12, 2005.

As a result, if the laser uses a 0.15-m diameter mirror, the power per steradian from the laser is roughly 10^{10}P W/sr. Since the laser's mirror is smaller than 1 m, the satellite's optics focus the laser light onto one pixel.

The intensity reaching the satellite from the laser (10^{10}P W/sr) is therefore equal to the intensity from a 1-m^2 piece of the Earth (100 W/sr) for a laser power P of 10^{-8} W. To ensure that the laser power overwhelms the reflected sunlight, we assume that the laser power should be 10 times the reflected light. By this estimate, a laser with a power P of 0.1 microwatts (μW) could dazzle the pixel on which the light from the laser mirror was imaged. Since each pixel corresponds to only a size of 1 m^2 on the ground, this is not useful to the dazzler since obscuring such a small area is unlikely to be useful.[3]

However, as noted in the text, the laser light is not perfectly focused onto one pixel, but is spread out over a larger part of the detector by several mechanisms. For this analysis, we assume that reflections off surfaces within the satellite's optical system have been eliminated. Instead we consider light that is diffracted by the satellite's optics and light that is spread over the focal plane around a central peak by imperfections in the optical system. We write the intensity at a distance d from the central peak as I(d) = A(d)I(0).

Following the discussion above, we assume that a pixel at a distance d from the central peak will be dazzled if the laser intensity reaching that pixel, which is smaller than the intensity at the central peak by the attenuation factor A(d), is ten times greater than the light reaching that pixel from the ground. As a result, the laser power P_d required to dazzle pixels out to a distance d is roughly 0.1/A(d) μW.

Due to diffraction by the satellite's circular mirror, the image of 1 m2 on the ground is focused to a spot on the detector with a diameter of roughly 1.22 fλ /D_s, where D_s is the diameter of the satellite's mirror, f is its focal length, and λ is the laser wavelength; this length on the detector therefore corresponds to roughly 1 m on

3. If the intensity on a pixel is high enough to saturate it, the electrons it generates can overflow the pixel's storage bin. Depending on how the array is designed, these electrons can spill over onto neighboring pixels in a process called blooming. However, there appear to be ways to design the array so that electrons that spill over are carried away from the detector rather than affecting neighboring pixels.

the ground. The central spot is surrounded by concentric rings that make up the diffraction pattern. The central peak contains roughly 84% of the light from the image;[4] the remaining 16% is spread into the diffraction rings. The diffraction pattern (including only the diffraction effect of the finite diameter of the satellite's mirror, but not the diffraction from the satellite's support structure and secondary mirror, which may comprise an additional few percent of the peak) can be approximated as a Fraunhofer diffraction pattern from a circular aperture, and the spacing and intensity of the rings can be calculated (see Appendix C to Section 11).

The maximum intensity of the first ring is 0.018 times the intensity of the central peak. By the fourth ring, which corresponds to a distance of about 4 meters on the ground, the intensity has dropped to less than 10–3 of its value at the central peak. Far from the central peak, the intensity of the maxima in the diffraction pattern falls off as one over the cube of the distance from the central peak (Appendix C to Section 11).

Techniques exist for suppressing the intensity of the diffraction rings by modifying the optics. This process, called apodizing, may lead to some broadening of the central peak, which reduces the resolution of the system. It may also reduce the total amount of light that gets to the sensor, which may degrade the image. Whether it makes sense to reduce the intensity of the diffraction rings depends on the level of stray light in the system that comes from other sources.

The intensity of stray light from imperfections in the satellite's optical system appears to decrease more slowly with distance from the central peak than the diffraction peaks, and will therefore be the dominant source of light far from the central peak. Published studies of the Hubble Space Telescope and follow-on systems suggest that the intensity of stray light may decrease roughly as one over the square of the distance from the central peak.[5]

---

4. Max Born and Emil Wolf, Principles of Optics, 7th Edition (Cambridge, England: Cambridge University Press, 2003), 443.

5. John E. Krist, "WFPC2 Ghosts, Scatter, and PSF Field Dependence," August 9, 1995, http://www.stsci.edu/software/tinytim/tinytim.html, accessed December 15, 2004, and Pierre Bely et al., "NGST Optical Quality Guidelines," August 24, 2001, http://www.ngst.nasa.gov/ public/unconfigured/doc_0791/rev_04/ monograph7_v10.pdf, accessed December 15, 2004.

To estimate the laser power required to dazzle pixels out to a given distance from the central spot, we assume that A(d) = 10^{-3} at a distance d corresponding to 3 to 4 m on the earth, as it would be for a pure Fraunhofer diffraction pattern, and falls off as d^{-2} for larger d, corresponding to the falloff for scattering due to optical imperfections.

This estimate suggests that, under the assumptions made above, a 1-mW laser, operating at a wavelength of 1 μm and focused by a 0.15-m mirror, could dazzle a section of the satellite's detector corresponding to a ground image with a radius of about 10 m around the laser. One milliwatt is roughly the power of a standard laser pointer. Furthermore, under these assumptions, to increase the radius of the area that can be dazzled by a factor of 10, the laser power must increase by a factor of 100. Approximate results are given in Table 11.1.

Table 11.1. This table illustrates how the power of a ground-based laser needed to dazzle part of a satellite's detector increases with the size of the ground area corresponding to that section of the detector. This calculation assumes a laser operating at 1 μm focused by a 0.15-m mirror, and a satellite with 1-m resolution.

Radius of Ground Area Corresponding to Dazzled Region of Detector	Required Laser Power
10 m	1 mW
100 m	0.1 W
1 km	10 W
10 km	1 kW

The discussion in Appendix D to Section 11 shows that by the time the laser power is high enough to dazzle a large section of the detector, the central peak is bright enough to damage the detector in that area. We emphasize that the values in Table 11.1 are approximate and depend on the level of stray light in the optical system, and will differ for specific systems.

Appendix - 11 B

Angular Size, Resolution, and Beam Divergence

Angular Size and Resolution

The apparent size of an object can be expressed in angular units and is called its angular size. Two sets of units are commonly used. A circle consists of 2π radians (r) or 360 degrees (°). There are 60 arcminutes (denoted by ') in each degree of arc, and 60 arcseconds (denoted by ") in each arcminute. Angular sizes are related to physical sizes by the object's distance: the angle subtended by an object is proportional to its physical size and inversely proportional to its distance. For example, a bicycle viewed from 100 m subtends roughly the same angle as a bus viewed from 400 m. It turns out that the Sun, at a distance of 1.5×10^8 km from Earth, subtends the same angle as the Moon, at a distance of around 3.85×10^4 km, because the ratio of their physical sizes happens to be the same as the ratio of their distances from the Earth.

The angular size $\Delta\theta$ of an object (for small angles) is given in radians by

$$\Delta\theta l/d \tag{11.1}$$

where l is its physical size and d is its distance. For l in meters and d in kilometers, $\Delta\theta$ is given in units of arcseconds by

$$\Delta\theta = 20.6^{\pi}\left[\frac{l/10}{d/100}\right] \tag{11.2}$$

The angular resolution of a telescope depends on the diameter D of the telescopic lens or mirror and the wavelength λ of the radiation the telescope gathers. It is given in radians by

$$\Delta\theta = 1.22\ \lambda/D \tag{11.3}$$

or in arcseconds by

$$\Delta\theta = 0.14'' \left[\frac{\lambda/550}{D}\right] \tag{11.4}$$

where λ is measured in nanometers and D in meters.

The resolution improves as the wavelength gets shorter (visible light has a shorter wavelength than radio waves, and ultraviolet light has a shorter wavelength than visible light) and as the diameter gets larger. For example, an optical telescope (which gathers visible light) with a diameter of 10 m has an angular resolution of roughly 0.1 microradian (μr) or 0.02, whereas a radio telescope of the same size has a resolution of roughly 0.1 milliradian (mr) or 20—making it worse by a factor of 1,000.

The angular resolution indicates how far apart two objects have to be in order to be seen as separate objects rather than one object. A telescope with angular resolution of 20 μr (4) could not distinguish two stars that are separated by 10 μr (2), but a telescope with an angular resolution of 5 μr (1) could. The angular resolution also indicates how much detail the telescope can observe about an object at a specific distance.

For an object at 500 km, a telescope with a resolution of 2 μr (0.4) could see detail on the scale of 1 m; one with a resolution of 20 μr (4) could observe details on the scale of only 10 m. For an imaging satellite, the size of objects it can see on the ground depends on the altitude of the satellite and on the satellite's optics. The image of the ground is focused onto the focal plane, where the sensor is mounted. The imaging system is usually designed so that the physical size of the smallest object that can be imaged by the satellite (called the resolution element, given by the angular resolution multiplied by the effective focal length of the optics) matches the pixel size of the sensor, so that the resolution element falls onto one or a few pixels.

Beam Divergence

The divergence of a beam transmitted by a telescope is closely related to the angular resolution of the telescope when it is used to observe an object. A telescope of a given diameter transmits a beam of approximately $\Delta\theta = 1.22\lambda/D$, where the size of the beam is given in radians and effects of the atmosphere are neglected. The size of the beam, l, at a given distance away can be found by multiplying the beam size by the distance, d, so that $l = \Delta\theta \times d$. For example, a beam of light with wavelength 1 μm, being focused by a telescope of 1-m diameter, produces a beam of diameter 0.85 m at a distance of 700 km.

Appendix - 11 C

Fraunhofer Diffraction Pattern

The intensity I of the Fraunhofer diffraction pattern, assuming light of wavelength λ passing through a circular aperture of diameter D, is[1]

$$V = \sqrt{GM_e\left(\frac{2}{h+R_e} - \frac{1}{a}\right)} = \sqrt{GM_e\left(\frac{2}{r} - \frac{1}{a}\right)} \tag{11.5}$$

where $I(0)$ is the intensity at the central maximum, J_1 is a Bessel function of the first kind of order 1, f is the focal length of the satellite's telescope, and d is the distance from the optic axis on the focal plane, where the detector is located. The pattern is axially symmetric around the central maximum, which lies on the optic axis. The maxima and minima of I/I(0) are given by the condition[2]

$$0 = \frac{d}{dx}\left(\frac{I}{I(0)}\right) = 8\left(\frac{J_1}{x}\right)\left(\frac{J_1'}{x} - \frac{J_1}{x^2}\right) = -\frac{8}{x^2}J_1J_2 \tag{11.6}$$

The minima occur when $J_1 = 0$ since this makes $I/I(0) = 0$. When I is nonzero, J1 must be non-zero; in this case Equation 11.6 requires $J_2 = 0$, and this is the condition for the secondary maxima. For large m, the mth zero of J_2, and therefore the mth secondary maxima of I/I(0), occurs approximately at $x_m = (m + \frac{3}{4})\pi$,

1 "Fraunhofer Diffraction-Circular Aperture," http://scienceworld.wolfram.com/ physics/ FraunhoferDiffractionCircularAperture.html, accessed February 6, 2005.

2 The final equality uses Eq. 9.1.27 from F.W.J. Oliver, "Bessel Functions of Integer Order," in Handbook of Mathematical Functions, ed. Milton Abramowitz and Irene A. Stegun (Washington, DC: Government Printing Office, 1972), 361, online at http://jove.prohosting. com/ ~skripty/page_355. htm, accessed February 6, 2005.

where m = 1 refers to the first secondary maximum.[3] The value of *d* at the secondary maxima is therefore linear in m. For large x, the maximum values of J_1 have the form[4]

$$J_1(x) \approx \sqrt{\frac{2}{\pi x}} \tag{11.7}$$

so that the intensity at the secondary maxima has the form

$$\frac{I}{I(0)} \approx \frac{8}{\pi}\frac{1}{x^3} \tag{11.8}$$

Table 11.2 lists the locations of the secondary maxima of the diffraction pattern and the corresponding intensities.

Table 11.2. This table shows the locations of the secondary maxima of the diffraction pattern and the intensity of those peaks relative to the intensity of the central maximum, I /I (0). Here xm is the argument of the Bessel function at the mth maximum, d/ d0 is the distance (on the detector) of the maximum from the central maximum in units of $d_0 = 1.22\lambda f/D$, which is the distance on the detector corresponding to the angular resolution of the satellite's optics (see Appendix B to Section 11). For the case considered here, d_0 corresponds to about 1 m on the ground.

m	x_m	d/d_0	$I/I(0)$
1 5.	14	1.3	0.018
2 8.	42	2.2	0.0042
3	11.6	3.0	0.0016
4	14.8	3.9	0.00078
5	18.0	4.7	0.00044
6	21.1	5.5	0.00027
7	24.3	6.3	0.00018
8	27.4	7.2	0.00012
9	30.6	8.0	0.000089

3. Oliver, 371, Eq. 9.5.12.

4.. Oliver, 364, Eq. 9.2.1.

Appendix - 11 D

Power Estimate for Laser Blinding

To develop a rough estimate of the laser power required to damage pixels in a satellite's detector, we assume the damage threshold for a silicon detector is 10^6 J/m^2 of incident energy delivered in less than 10^{-4} seconds. The damage threshold depends on the detector material; the value for silicon appears to be high compared with other materials.[1] A ground-based laser of power P operating at wavelength λ and using a mirror of diameter D_L spreads the laser light over a disk with diameter of roughly $1.22\lambda R/D_L$ at a distance R, giving an intensity at that distance of

$$I = \frac{P}{\pi\left(\frac{1.22\lambda R}{2D_L}\right)^2} = \frac{4D_L^2 P}{\pi(1.22\lambda R)^2} \qquad (11.9)$$

As in the discussion of dazzling in Appendix A to Section 11, the laser light collected by the satellite's optical system is assumed to be focused onto one pixel. The optical system will concentrate the light by a factor C, which is roughly the ratio of the area of the satellite's telescope to the area on the detector onto which this light is focused. If D_S is the diameter of the satellite's telescope and f is its focal length, the ratio of these areas is D_S *2/[1.22λ f/ D_S]2. For $D_S = 1$ m, $f = 2$ m, and $\lambda = 1$ μm*, this ratio is greater than 1011. Over a time Δt, the energy incident on the pixel is then

1 Silicon has a damage threshold at an irradiation time of 10–4 seconds of about 1010 W/m2, for a total energy deposited of 106 J/m2. The damage threshold is similar for a range of incident wavelengths, from 0.69 μm to 10.6 μm. Other common detector materials have lower damage thresholds at this timescale: InSb and HgCdTe thresholds are around 5 × 104 J/m2. See F. Bartoli, L. Esterowitz, M. Kruer, and R. Allen, "Irreversible laser damage in IR detector materials," Applied Optics 16 (November 1977): 2934–2937.

$$\frac{Energy}{area} = CI\Delta t = \frac{4C\,D_L^2 P\Delta t}{\pi(1.22\lambda R)^2} \qquad (11.10)$$

The laser power required to damage the pixel on which this light is focused is found by setting this expression equal to the energy per area needed to damage the detector, taken here to be 10^6 J/m^2. Assuming the distance to the satellite is 800 km, the laser wavelength is 1 μm, and the concentration factor C is 1010, this equation gives a condition for achieving the damage threshold at the pixel onto which the laser is imaged

$$d_L^2 P\,\Delta t \approx 10^{-4}\,Jm^2 \qquad (11.12)$$

The time it takes for the satellite to pass over a 1-m^2 area of the Earth is roughly 10^{-4} s, so this will be the time the detector will have to collect light from that small ground area. Using this time for Δt, a CW laser with a mirror diameter of 0.15 m and a power of 40 W would therefore be able to meet the damage criteria in Equation 11.11. Using a 1-m diameter mirror with adaptive optics, the required power could be reduced to about 1 W.

For a pulsed laser, Δt is taken as the length of a pulse (as long as it is less that 10^{-4} s) and P as the peak power. The quantity PΔt is then roughly the total energy per pulse. For a mirror diameter of 0.15 m, a laser that produces pulses with energy greater than about 4 mJ can satisfy the damage criteria in Equation 11.11, assuming the pulse width is less than 10^{-4} s. For a 1-m mirror, a laser producing pulses with an energy of 0.1 mJ will satisfy Equation 11.11. If the pulse width and the time between pulses is short enough that several pulses are produced in 10^{-4} s, then $P\Delta t$ is the sum of the energies of all those pulses.

These numbers give a rough estimate of the power levels required for damaging a few pixels on a detector.

As in the discussion of dazzling (Appendix A to Section 11), we note that the optical system will spread some of the laser light over a larger part of the detector. Under the same assumptions as in the discussion of dazzling, the intensity of the laser light at a distance of about 10 pixels from the central peak of the diffraction pattern, corresponding to a ground distance of about 10 m, would be roughly 10^{-4} times that maximum intensity. Delivering enough light to a pixel at that distance to damage it would therefore require a laser power 104 times larger than needed just to damage the central pixel, or 400 kW for the CW case with a 0.15-m mirror, and 10 kW for a 1-m mirror. While CW lasers with powers of 10 kW are commercially available, 400 kW lasers are not. Assuming, as in Appendix A to Section 11, that the intensity of the stray light in the detector falls off as one over the square of the distance from the central peak, increasing by a factor of ten the damaged area on the detector requires increasing the laser power by a factor of 100 (see Table 11.3).

For mirror sizes of 0.15 m and 1 m, a pulsed laser able to produce pulses with energy greater than 40 J and 1 J, respectively, with pulse widths less than 10^{-4} s, would be able to damage the detector out to a distance of about 10 pixels from the central peak by the criteria used here. Commercial industrial lasers with pulse energies of tens of joules are sold for applications such as welding and drilling.[2] Using a 1-m mirror, a laser capable of producing 100 J pulses could damage a detector out to a distance of about 100 pixels from the central peak, corresponding to a ground distance of about 100 m, under the assumptions used here.

These results suggest that lasers with commercial-level power might damage the detectors of high-resolution imaging satellites over areas corresponding to tens of meters on the ground.

The power required for damaging the detectors of actual satellites depends on details of the detector and optical system, and may differ, perhaps significantly, from these estimates.

Table 11.3. This table illustrates how the required power (for a CW laser) or pulse energy (for a pulsed laser) needed to damage part of a satellite's detector depends on the size of the ground area corresponding to that section of the sensor, and on the size of the laser's mirror. This calculation assumes a laser operating at 1 μm focused by either a 0.15-m or 1-m mirror, and a satellite with 1-m resolution.

2 See, for example, U.S. Laser Corporation, "High Power Pulsed Nd:YAG Laser," http:// www. uslasercorp.com/pulsedspecs.html, accessed January 14, 2005.

Laser mirror diameter D_L = 0.15 m		
Ground distance corresponding to detector damage	Power (CW laser)	Pulse energy (Pulsed laser)
1 m	40 W	4 mJ
10 m	0.4 MW	40 J

Laser mirror diameter D_L = 1 m		
Ground distance corresponding to detector damage	Power (CW laser)	Pulse energy (Pulsed laser)
1 m	1 W	0.1 mJ
10 m	10 kW	1 J
100 m	1 MW	100 J

Section - 12

Topics in Interfering with Satellites

This section discusses in more detail three topics related to interfering with satellites: space-based anti-satellite weapons (ASATs), including space mines; simple ground-based pellet ASATs, which could potentially be deployed by nonspacefaring countries; and attacks on and back-up alternatives to the global positioning system and reconnaissance satellites.

Space - Based Asats and Space Mines

Space-based ASATs would be launched into orbit in advance of use and would be continuously or occasionally in range of a target satellite.[1] Space mines are space-based ASATs, but the term has no precise definition. Some authors use the term loosely, others define it specifically, but these definitions differ from author to author. This report avoids the term.

Space-based ASATs could be launched at the beginning of a crisis or placed in orbit in anticipation of a potential future conflict. If a country is concerned about its ability to launch promptly due to weather or other factors, or if it is concerned about the potential suppression of its launch capabilities during a conflict, it could address these reliability concerns by positioning its ASATs in space ahead of time. On the other hand, if these objects are identified as ASATs and are seen as threatening, they would be vulnerable to attack. Moreover, satellite reliability degrades over

1. We do not include in this category those satellites that serve some other primary function (e.g., ballistic missile defense or inspecting other satellites) and have an inherent ASAT capability.

time, so the owner will have decreasing confidence in space-based ASATs after they have been placed in orbit.

Such a weapon is likely to be stationed in one of four ways. It could be coorbital with the target satellite, mimicking its stationkeeping maneuvers and keeping within a fixed distance; this is referred to as a trailing ASAT. It could attach itself to its target, eliminating the need to track the target satellite and carry fuel for stationkeeping; such satellites have been dubbed parasitic.[2] (Coming into physical contact with another satellite, however, is likely to be considered unlawful or provocative.) The satellite could be placed in a distant part of the same orbit, which would require it to maneuver to approach and attack the target. Finally, it could be placed in a crossing orbit in the same orbital plane as its target or in a different plane.

In principle, a space-based ASAT could cause temporary or permanent damage to its target using many of the methods discussed in Section 11. We discuss below which means of attack would be possible and practical for space-based ASATs and compare their potential performance with that of ground-based ASATs. Some deployment options may be better suited than others to a particular method of attack. For example, for a kinetic energy attack, an ASAT in a crossing orbit could take advantage of its high speed relative to the target satellite. But a kinetic attack from a trailing ASAT would require it to explode near the satellite or shoot pellets at it, either of which would impart less energy. On the other hand, a trailing ASAT could attack almost instantaneously, whereas a crossing ASAT would need to wait until it was in the proper position. An ASAT in a crossing orbit might, however, be capable of attacking multiple satellites, whereas a trailing or attached ASAT could be used against only one.

Covert Space-Based ASATs

The choice of orbit affects the ability of the owner of the space-based ASAT to keep its existence or its purpose covert. ASATs in crossing orbits would be less suspicious than trailing or co-orbital ASATs and might also be less readily detected.

2. See Gregory Kulacki and David Wright, "A Military Intelligence Failure? The Case of the Parasite Satellite," http://www.ucsusa.org/global_security/china/page.cfm?pageID=1479, accessed February 9, 2005.

If an ASAT was not deployed covertly, the owner of the targeted satellite might take action, seeking to make an international issue out of the deployment, particularly if it could determine the owner of the ASAT. The satellite owner or the wider international community could demand that the ASAT be removed or could assign responsibility if it was used, thereby legitimizing retaliation. Equally important, noncovert deployment would remove the element of strategic surprise from an attack and give the targeted country time to develop a contingency plan to compensate for missing satellites. The satellite owner might also decide to preemptively attack the space-based ASAT. However, preemptive attack on parasitic and trailing ASATs is especially difficult because of the proximity of these ASATs to the targeted satellite.

Thus the ASAT's owner is likely to want to keep its existence or purpose covert. However, the owner could not assume that the ASAT would remain covert and would need to factor that into the deployment decision. There may, of course, be situations in which the ASAT's owner would want other countries to be aware of the ASAT's existence in order to send a political signal.

A country seeking to deploy a covert space-based ASAT might attempt to prevent detection of its launch. For example, a small ASAT could be launched along with a legitimate satellite. In addition, the ASAT's owner would try to prevent its detection once in orbit or might conceal its purpose by disguising it.

Currently, most satellites are launched into orbit from a small number of fixed launch pads,[3] and launches are announced in advance. Such space launches are readily observable from the ground and any country that wanted to monitor these launch sites could do so. However, small payloads can be launched from aircraft and smaller ground- and sea-based facilities. While the United States and possibly Russia could detect such launches with their existing early warning satellites, no other countries currently have that capability. However, the Russian system has never provided global coverage, since it is designed to detect launches from the United States.[4] Thus the United States might be able to covertly launch

3. The Sea Launch system is a floating platform used to launch rockets from near the equator. However, it prepares the launcher and loads the satellite in Long Beach, California, and is unlikely to go unnoticed.

4. Pavel Podvig, "History and the Current Status of the Russian Early Warning

space-based ASATs from aircraft or small seabased launch facilities. This possibility only applies to the United States, since all space launches by other countries would be detected by the U.S. early warning system.

Rather than trying to hide the launch itself, a country could attempt to hide the deployment of a small ASAT by placing it on the same launcher as a legitimate satellite and announcing only the deployment of the satellite. If the ASAT was small enough and the bus did not maneuver to place it in orbit, its deployment might go undetected.

Again, such deployment is unlikely to go undetected by the United States, which maintains an extensive Space Surveillance Network (SSN). The SSN consists of optical sensors and radars that track the roughly 8,500 objects 10 centimeters or larger that are orbiting Earth (including some 600 operational satellites, 1,300 rocket bodies, and 6,600 inactive satellites or other space debris). The SSN charts the position of these objects and plots their anticipated orbital paths. When the U.S. early warning satellites detect a rocket launch, the SSN detects its "associated objects," such as debris and deployed satellites. Thus, the United States is likely to detect deployment of even a small space-based ASAT, unless the ASAT uses techniques to reduce its optical and radar signatures.[5] Other countries currently have more modest space detection and tracking capabilities, so that it is possible the United States could deploy a small space-based ASAT without detection.

Even if the launch or deployment of a space-based ASAT were not observed, it might still be detected in orbit—sooner or later. However, if the ASAT did not maneuver to place itself in a new orbit, it could appear to be a piece of debris. The SSN does not have the capability for real-time data analysis and observes the operational satellites more frequently than it does debris. Nevertheless, once the ASAT maneuvered, the SSN would likely identify it as a satellite. All the other space-faring nations (Russia, China, Japan,

---

System," Science and Global Security 10 (2002): 21-60, http://www.russianforces.org/podvig/eng/publications/ sprn/20020628ew/index.shtml, accessed January 5, 2005.

5. "Misty," a CIA satellite launched in 1990, is thought to have used an inflatable reflective shield to evade detection by Soviet surveillance for many years. Robert Windrem, "A Spy Satellite's Rise...And Faked Fall," July 12, 2001, http://msnbc.msn.com/id/3077830/, accessed February 9, 2005.

India, Israel, Ukraine and many of those of the European Union) have sufficient surveillance capability to monitor their own satellites and to detect trailing satellites in low earth orbits and possibly in geosynchronous orbits. It is less likely that these countries would survey objects in an orbit other than the ones their satellites occupy, so it is possible that a U.S. space-based ASAT in a crossing orbit could remain covert, at least for some time.

However, as noted above, a space-based ASAT could be designed to limit its optical and radar signatures. Moreover, as satellite miniaturization techniques continue to improve, space-based ASATs with dimensions of a few tens of centimeters or smaller will be feasible. In the face of such developments, detection of space-based ASATs may become difficult even for the United States.

In addition to detecting the physical presence of a space-based ASAT, it may be possible to discover an ASAT by detecting its communications with the ground, although these may be short and infrequent.

Even if a space-based ASAT was detected and tracked and determined to be a satellite, its purpose could remain covert. A space-based ASAT could be disguised as a legitimate satellite. However, the ASAT's orbit would give a clue to its purpose, especially if it closely trailed another satellite. A satellite in a crossing orbit would be less likely to raise suspicions.

In sum, no country can assume it would be able to detect the ASATs deployed by other countries. At the same time, no country can assume that its deployment of space-based ASATs would remain covert, not even the United States. However, this situation may change in the future, given current trends in satellite miniaturization and techniques to reduce satellite signatures.

Space-Based ASATs vs. Ground-Based ASATs

Depending on the means of attack, space-based ASATs and ground-based ASATs have relative advantages and disadvantages. Section 11 considers the suitability of ground- and space-basing for the various methods of interference considered there; Table 12.1 summarizes those results. As discussed in Section 11, jamming and dazzling are not well suited to space basing, since they would require essentially constant maneuvering to be in position to attack a target satellite. For the foreseeable future, using space-based lasers to damage the structure of a satellite, as opposed to its sensor, is not technically feasible. On the other hand, high power microwave attacks are not well suited to delivery from Earth, but can be delivered from space-based HPM generators.

Table 12.1. The Xs indicate that the method of interference is well suited to basing on the ground or in space.

Ground-based	Space-based	
Uplink jamming	X	
Downlink jamming	X	
Dazzling	X	
Partial blinding	X	X
High power microwaves	X	
Laser damage	X	
Kinetic energy	X	X
Nuclear weapon	X	

The analysis in Section 11 shows that it is feasible to partially blind a satellite's optical sensor using either a ground-based laser or a modest sized spacebased laser in a crossing orbit, although in either case the number of pixels affected would likely be small. Kinetic kill attacks could also be conducted using either ground- or space-based ASATs.

This subsection examines in more detail the relative advantages and disadvantages of using ground- and space-based ASATs designed to partially blind a satellite or destroy it by kinetic means. For these two types of attacks, we assess the relative ability of space- and ground-based ASATs to

- undergo covert development, testing, and deployment and thus deny strategic warning to an adversary
- work effectively and reliably
- deliver an attack on multiple satellites in a short period of time and thus limit tactical warning.

We also consider the ability of the adversary to counter an attack. Relative cost is not discussed, although in practice this will be an important consideration.

Covert Development and Testing. Although the development and testing of ASATs—whether space- or ground-based—would not be as provocative as ASAT deployment, it could nonetheless raise objections within the international community and warn potential adversaries of forthcoming deployment. Hence a country developing such weapons would likely prefer to do so covertly.

A space-based ASAT designed to partially blind a satellite could not be fully tested unless it was placed in orbit, which would pose a risk of detection. However, the purpose of the ASAT could be kept covert, and it is unlikely that others could detect the actual testing. A country could test ground-based lasers against its own satellites—either those at the end of their lifetime or satellites designed for this purpose—with little risk of detection.

Homing kinetic energy ASATs would impact their target at closing speeds of roughly 7 to 14 km/sec. The ability to directly impact a moving target at such high closing speeds can be assessed only by intercept testing. A country would find it difficult to conduct a covert intercept test of either ground- or space-based kinetic energy ASATs, since the resulting collision would generate significant debris and eliminate the original satellite. The United States would certainly be able to detect such a collision relatively quickly. Other space-faring countries would also be able to detect it, sooner or later.

On the other hand, a country could conduct the required tests in the context of developing an exo-atmospheric hit-to-kill missile defense system, since the closing speeds are comparable. In the intercept tests of the U.S. Ground- Based Midcourse Defense (GMD) to date, the time and details of the attack were known in advance, the trajectory of the target warhead was known by the defense, and there were no decoys. While these conditions are not appropriate to test a realistic attack by a ballistic missile warhead, they are appropriate to test intercepting a satellite in low earth orbit.

Covert Deployment. As discussed above, a country may be able to covertly deploy space-based ASATs. This would be more feasible for ASATs in crossing orbits than for co-orbital ASATs. On the other hand, because ground-based lasers for partial blinding need not be enormous, they could be deployed covertly. And, while not covert, the deployment of interceptors as part of a ground-based midcourse missile defense provides a significant inherent ASAT capability against satellites at altitudes of up to several thousand kilometers, although this capability may not be widely recognized.[6]

Effectiveness. There is no reason to expect ground-based kinetic energy ASATs to be more or less effective than their space-based counterparts, whether homing interceptors in crossing orbits or co-orbital space-based ASATs that would explode near the target satellite or fire pellets at it. Additional analysis is required to assess whether the effectiveness of ground- and space-based laser ASATs designed to partially blind a target satellite differs significantly.

Reliability. A ground-based system might be more reliable than space-based ASATs, because it could be regularly maintained and upgraded. It may also be more feasible for a country to have back-up ground-based ASATs to use if the first ASAT, ground or space-based, fails.

Time to Deliver Attack. The ability to respond on the scale of minutes once an attack is ordered may not be essential for ASATs, in contrast to ground attack weapons that might seek to destroy mobile targets. However, a short response time might be useful in some situations.

Because of its proximity to the target satellite, a trailing kinetic energy ASAT could in principle attack rapidly once a decision was made to do so. This assumes, however, that the owner could communicate quickly with the ASAT, which may not be possible if the ASAT is not in view of a ground station or relay satellite to receive instructions. For the United States, which has ground stations worldwide, communication could be quick. A country with limited global presence could take considerably longer to send

6. See David Wright and Laura Grego, "Anti-Satellite Capabilities of Planned US Missile Defense Systems," December 9, 2002, http://www.ucsusa.org/global_security/ space_weapons/ page.cfm?pageID=1152, accessed January 17, 2005.

the signal to attack, although the signal may be simple enough that it could be sent from the country's diplomatic missions.

In contrast, an ASAT in a crossing orbit may require hours before it is in position to attack. Similarly, it could be hours before a ground-based ASAT was in the proper position to attack a satellite in low earth orbit. This time could be reduced if ground-based interceptors were positioned at various places around the globe, such as kinetic energy ASATs on aircraft. A ground-based kinetic energy ASAT would reach a satellite in low earth orbit in a matter of minutes, but it would take several hours to reach a satellite in geosynchronous orbit.

Perhaps more useful would be a capability to attack multiple satellites simultaneously or in a short period of time rather than over several hours or days. Such a rapid multiple attack could limit tactical warning for the targeted country and thus its ability to take action. A short attack time (the duration of the attack, once initiated) would not require a short response time (the time between a decision to attack and initiation of the attack).

If the attacker used trailing kinetic energy ASATs, these could attack numerous satellites essentially simultaneously, providing no tactical warning. A simultaneous attack would be more difficult using ASATs in crossing orbits, whose trajectories would need to be synchronized.

The ability to launch a multisatellite attack of short duration using groundbased ASATs would depend in part on how widely separated the satellites were at the time of attack and on whether the country could deploy ASATs at different locations.

Feasibility of Defense. Once an attack was under way, countering either a ground- or space-based kinetic energy ASAT would be difficult. Perhaps the most feasible approach would be to try to destroy the ASAT using a spacebased kinetic energy weapon. Doing so would require the ability to observe launches, which only the United States is capable of doing globally. However, for attacks against satellites in low earth orbit, the warning time would likely be too short to allow a counter attack. It may, in principle, be possible to intercept a kinetic energy ASAT attacking a geosynchronous satellite.

Once a ground-based blinding laser was attacking a satellite, its location could be determined and it could presumably be destroyed. A space-based laser in a crossing orbit would be vulnerable to attack by a kinetic energy ASAT. However, since partial blinding occurs quickly, this destruction would only prevent future attacks, not defend against the current one.

Simple Ground - Based Pellet (Non-Homing) Asat

Discussions of ASATs frequently refer to a low-tech method for attacking satellites in low earth orbit, which consists of a missile that does not home on the target satellite but lofts a large mass of sand or pellets into its path. Because the satellite in orbit is moving faster than 7 km/s, a collision with even a small particle can do severe damage to the satellite. Moreover, since the pellets are lofted into the path of the satellite and are not placed into orbit, they can be launched on relatively short-range missiles to attack satellites in low earth orbits.[7]

This subsection analyzes the probability that such a pellet ASAT successfully destroys a satellite. The results indicate that, in its simplest form, a pellet ASAT may have limited effectiveness. A country with sufficient technical capability could take various steps, discussed below, to increase the probability of intercept. Thus the technical capability assumed of the country using the ASAT must be clearly delineated. A simple pellet ASAT might be used by a country lacking the technical capability to develop an ASAT that would provide

higher confidence of success, such as a homing interceptor. (A homing interceptor might dispense a cloud of pellets to increase the probability of killing the satellite, but this would be a small cloud released shortly before the intercept, rather than a large cloud released much earlier.)

The ASAT considered here uses a missile fired roughly vertically to loft a cloud of pellets into the path of a satellite. Shortly after the missile burns out and stops accelerating, an

7. Since the pellets are not placed in orbit, they would fall back to Earth and not constitute orbital debris. If they struck a satellite, some of the particles produced could become long-lived debris, depending on their speed and direction of motion.

explosive charge or other mechanism disperses the pellets carried by the missile so that they form an expanding cloud (see Figure 12.1). The size of the cloud when it reaches the intended intercept point depends on the expansion speed of the pellets and the amount of time between the release of the pellets and when they reach the intercept point.[8] The path the cloud follows is controlled by aiming the missile, that is, controlling the missile's speed and direction at burnout. The center of mass of the cloud follows the same trajectory as would a simple warhead released from the missile.

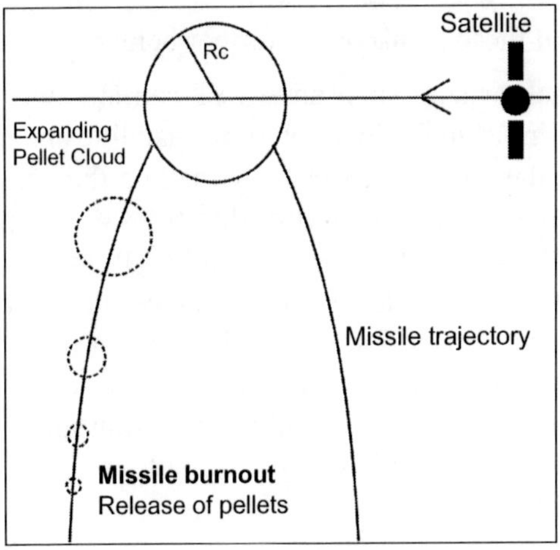

Figure 12.1. At burnout, a missile releases a cloud of pellets, which expands as it travels toward the intercept point, shown here at the apex of the trajectory, with a radius R_c. Timing errors can be minimized by arranging to have the cloud reach its maximum height at the altitude of the satellite's orbit.

The effectiveness of a pellet ASAT depends on the probability that one or more pellets strike the satellite and that those strikes disable the satellite. The probability that pellets strike the satellite depends on several factors:

- the accuracy with which the attacker can determine the satellite's trajectory, which determines how accurately

8. For the situation considered here, the pellet cloud expands for about 340 s from the time it is released after missile burnout until it reaches apogee at 600 km altitude. As a result, the pellets need a speed of only about 3 m/s to produce a cloud radius of 1 km at apogee.

the future position of the satellite can be predicted and an intercept location and time can be calculated

- the accuracy with which the attacker can deliver a missile payload to a specific point in space, i.e., the calculated intercept point
- the size of the cloud of pellets and the number of pellets in the cloud
- the size of the satellite, in particular, the cross-sectional area it presents to the pellet cloud.

Each of these is discussed further below.

A country using such an ASAT can attempt to compensate for uncertainties in the location of the satellite and in the accuracy of placing the cloud in its path by making the pellet cloud large, which increases the probability that the satellite passes through the cloud. However, for a given number of pellets, a larger cloud size means a lower density of particles and therefore a lower probability of impact. Increasing the number of particles requires either increasing the total mass the missile must lift into space, which is limited by the capability of the missile, or reducing the mass of each pellet, which reduces a pellet's ability to damage the satellite if it hits.

Even if one or more pellets hit a satellite, the satellite may not be disabled. While a satellite may present a relatively large area to the cloud of pellets, the portion of the satellite that must be hit to disable it may be much smaller than its total area. For example, a large part of a satellite's total area may be solar panels. Pellet hits might degrade the performance of the solar panels, but the panels may be able to sustain a number of hits before the damage disables the satellite.

Moreover, if the pellets are too small, they may not cause sufficient damage to disable the satellite. The ability of a pellet to damage a satellite depends on both its diameter and mass.[9] For this

9. For high-speed collisions, the ability of a pellet to penetrate a target, such as the outer wall of a satellite, increases with its size. Once the pellet hits the target, it continues to penetrate until the shock wave created by the impact at the front of the pellets travels to the rear of the pellet. See, for example, Stephen Remillard, "Debris Production in Hypervelocity Impact ASAT Engagements," Air Force Institute of Technology (DTIC # AD-A230-467), December 1990, 26.

reason, grains of sand may be too small to be effective, especially if the satellite uses simple shielding.[10] Shielding materials now available could protect sensitive parts of a satellite from strikes by particles with mass greater than 1g and diameter greater than 1 cm.[11] While shielding has the disadvantage of adding to the satellite mass, high-value satellites are likely to include some shielding against orbital debris. Such shielding would set a lower limit on the size of pellets required for a successful attack.

The Appendix to Section 12 describes a simple model to calculate the probability that, under various conditions, a satellite is hit by one or more pellets. For each set of conditions, the calculation gives the intercept probability as a function of cloud size. It considers the case of an attacker with a missile that can carry 500 kg of pellets to an altitude of 600 km. According to the "1/2 Rule" (Section 8), such a missile would be roughly comparable to a North Korean Nodong missile.[12]

Uncertainty in Satellite Position

The accuracy with which a country can determine the orbit of a satellite depends on the type and number of sensors it has to observe the satellite and on the software it has to calculate, based on its observations, the satellite's orbit and location at a future time. If the orbit remains essentially constant over time, the estimate of the orbit can be made more accurate over time by including data from additional observations. If the orbit changes on a relatively

---

10. For example, a piece of debris estimated to be 0.2 mm in size and traveling at 3 to 6 km/s chipped but did not penetrate a window of the space shuttle. Grains of sand typically have diameters of 0.05 to 2 mm and masses in the range of 0.1 to 10 milligrams (see Marina Theodoris, "Mass of a Grain of Sand," 2003, http://hypertextbook.com/facts/2003/ Marina Theodoris.shtml, accessed January 5, 2005).

11. See, for example, "3M Nextel Ceramic Fabric Offers Space Age Protection," 3M (98-0400- 5217-1) 1997, http://www.3m.com/market/industrial/ceramics/pdfs/CeramicFabric.pdf, accessed December 5, 2005, and International Space Information Service, Technical Report on Space Debris, UN (A/AC.105/720) 1999, Section III.B.1, http://www.oosa.unvienna.org/isis/pub/ sdtechrep1/sect03b1.html, accessed December 5, 2005.

12. The Nodong missile is believed to have a maximum range of about 1,300 km with a 700 kg payload. Since the payload would need to include the structure of the stage containing the pellets and the mechanism to distribute them, the actual mass of pellets that such a missile could lift to this altitude would probably be less than 500 kg.

short time scale due to atmospheric drag, stationkeeping, or other maneuvers, additional observations will provide no help in refining the estimate of the orbit.

At the simplest level, telescopes can be used to measure a satellite's angular position in the sky. But telescopes cannot determine the distance to an object and therefore cannot directly measure the altitude of the satellite at any point. Instead, measuring the period of the orbit, and the satellite's position and angular speed at various points on the orbit provides the information necessary to estimate the shape and orientation of the orbit. Since the orbit will in general not be circular, the altitude of the satellite varies with its position on the orbit. A country may have difficulty collecting sufficient data to determine the orbit accurately for several reasons. For example, it may not have the ability to collect information from locations distributed around the world.[13] Since optical measurements require seeing sunlight reflected off the satellite, the satellite will be in the proper position to be seen from a given location for only relatively short periods of time during the day, which limits the data that an observer can collect as the satellite passes over. The observer may also not have the equipment and the ability to make highly accurate measurements. And, as noted above, the satellite may be maneuvering.

It is relatively difficult to determine the satellite's altitude at a given point on its orbit to high accuracy using measurements of this type. As shown below, even uncertainties in altitude of a few kilometers out of a total orbital radius of 7,000 km—corresponding to uncertainties of a few hundredths of a percent—are enough to significantly reduce the effectiveness of an ASAT.

Other types of measurements, if available, could reduce the uncertainty. If a country had a radar system that could detect the satellite to be attacked, it could determine the distance to the satellite accurately using the travel time of a radar pulse to the satellite and back to the radar. However, a country such as North Korea may have difficulty acquiring a radar with this capability.

13. Allen Thomson has suggested that countries might take optical measurements from its embassies around the world as a way of collecting global data. See Allen Thomson, "Satellite Vulnerability: A Post-Cold War Issue?" Space Policy 11 (1995) 19–30, http://www.fas.org/ spp/ eprint/at_sp.htm, accessed December 5, 2005.

Laser radar, which determines the range in the same way using light pulses, may be more feasible.

Inaccuracy in Positioning the Pellet Cloud

The inaccuracy in aiming the center of the cloud at a particular location is caused by errors in controlling the burnout speed of the missile (so-called guidance and control errors). These can be estimated from the accuracy with which the missile can deliver a warhead to a ground target. This accuracy is quantified by a circular error probability (CEP), which is the radius of a circle that includes half of the impact points of a large number of warheads fired at the same target. A country with relatively low technical sophistication will not be able to control the burnout speed accurately, which will lead to a relatively large CEP.

The CEP of the Nodong missile at its maximum range is estimated to be several kilometers. Assuming the guidance and control errors are a significant component of the total CEP,[14] this implies that the burnout velocity can be controlled to a few meters per second (compared to the burnout velocity of the Nodong of about 3 km/s). The analysis in this report assumes this level of technology.

If the pellet cloud is fired vertically to an altitude of 600 km, an error in the horizontal component of velocity of ±3 m/s leads to an error in the horizontal position of the center of the cloud of roughly ±1 km. Similarly, an error of ±3 m/s in the vertical velocity leads to an uncertainty in the maximum altitude, or apogee, of the missile trajectory of roughly the same amount. Moreover, these errors introduce an uncertainty into the time to reach apogee of several tenths of a second.

The CEP describes the spread of missile impact points. In general, that pattern of impact points will not be centered at the aim point of the missile due to systematic errors in guiding the missile, which affect all launches the same way, rather than shot-to-shot errors, which affect each launch differently. The distance between the aim point and the center of the impact pattern is called the bias. A large bias can increase the distance by which

14. Unpredictable atmospheric forces during reentry will contribute to the CEP when attacking targets on the ground, but not targets in space.

the pellet cloud misses its aim point. Since the CEP and the bias are determined statistically, a country that has done relatively few flight tests of the missile may have little information about the CEP or bias of the missile.

A country with sufficient technical ability could more accurately position the pellet cloud in space by taking measures to increase the accuracy of its missiles. One possibility is to add a small maneuvering bus to the missile that uses a GPS receiver and small thrusters to reduce the guidance and control errors.

Timing Errors

In order for an intercept to occur, the satellite and pellets must be at the same place at the same time. Errors in timing can arise from uncertainties either in predicting when the satellite will arrive at a certain point on its orbit or in delivering the pellet cloud to a given location at the right time.

A significant timing error can be tolerated if the satellite passes directly over the ASAT launch site, so that the ASAT is fired vertically and the cloud has no horizontal velocity. The ASAT is then fired so that the intercept point is at the apogee of the pellet cloud's trajectory. Since at apogee the cloud stops moving vertically and begins to fall, it remains in the satellite's path for tens of seconds.[15]

However, if the satellite's orbit does not pass directly over the ASAT launch site, the ASAT will need to travel some distance horizontally as it travels vertically to the proper altitude. As a result, the pellet cloud will have a horizontal component of speed V_h when it reaches the proper altitude, and a timing error of Δt will result in a position error for the cloud of $\Delta t \times V_h$.

Since the time from launch until the pellet cloud reaches an altitude of 600 km is about 440 s for a missile like the Nodong, the horizontal speed of the pellet cloud would be about 225 m/s for each 100 km of horizontal distance that the ASAT needs to travel from its launch site to the intercept point. Assuming the total timing uncertainty from all sources is 0.5 s, the cloud's

15. This possibility was suggested by Richard Garwin. At 600 km altitude, the cloud stays within 1 km of its apogee for 31 seconds, within 0.5 km of its apogee for 22 seconds, and within 0.25 km of apogee for 15 seconds.

horizontal speed would lead to a position uncertainty of about 110 m for each 100 km of horizontal distance. As a result, it would be advantageous for the country using this ASAT to launch its attack at the time the satellite passes roughly overhead or to launch the ASAT from a mobile launcher that could move directly under the satellite's path. Since the ASAT is launched on relatively short-range missiles, such mobility is possible. The Nodong missile, for example, is designed for use on a mobile launcher.

Results of the Calculation

The simple model described in the Appendix to Section 12 estimates the effectiveness of an ASAT of this type and describes the important parameters. The analysis considers a missile that can carry 500 kg of pellets to an altitude of 600 km. It uses the following parameter values for the base case and then considers variations around them:

- The uncertainty in the location of the pellet cloud. The base case assumes an uncertainty of 1 km, which corresponds roughly to the accuracy of a Nodong missile, as discussed above.[16]
- The uncertainty in the location of the satellite. The base case assumes that the attacker can determine the horizontal position of the satellite to about 0.1 km, but can predict the satellite's altitude at the planned intercept point only to within 1 km.[17]
- The frontal area of the satellite in which the collision with a pellet could disable the satellite. The base case assumes an area of 10 m^2.
- The number of pellets in the cloud. The base case assumes the cloud contains 500,000 pellets, each weighing 1 g, distributed uniformly within a spherical region. A 1-g spherical pellet made of aluminum has a diameter of 0.9 cm.

16. In particular, the probability distribution of the location of the pellet cloud at apogee is given by a Gaussian distribution with a standard deviation of 1 km. As a result, the center of the cloud falls within a circle of radius 1 km about 40% of the time.

17. In particular, the location of the satellite is given by a two-dimensional Gaussian distribution with standard deviations of 0.1 and 1 km in the horizontal and vertical directions.

The calculation varies the size of the pellet cloud to find the size that gives the maximum value of the intercept probability for each set of parameter values. The pellet cloud is assumed to be spherical, as discussed in the Appendix to Section 12.

The values given here are the probabilities that the vulnerable area of the satellite is struck by at least one pellet. The analysis does not address the issue of whether the pellets that strike the satellite will disable it. For the base case, the calculation shows the probability of the satellite being hit by at least one pellet is less than 30%, and the probability it would be struck by at least two pellets is 10%. The optimum radius of the pellet cloud in this case is about 1.4 km.

If the attacker can reduce his uncertainty in the altitude of the satellite from 1 km to a few tenths of a kilometer, the probability of pellet impact increases to about 35%. In this case, the optimum cloud radius is still well over 1 km since the inaccuracy in the placement of the cloud is still large. On the other hand, if the altitude uncertainty increases to 2 km, the probability of at least one pellet impact decreases to less than 20%. If it is as large as 5 km, the intercept probability is well under 10%.

If the missile accuracy is improved so that it can deliver the pellet cloud with an accuracy of 0.5 km rather than 1 km, the intercept probability of the base case increase to about 45%. If the accuracy is instead 1.5 km, the probability drops to under 20%.

Increasing the number of pellets by a factor of two, from 500,000 to 1 million increases the intercept probability of the base case to slightly over 40%. Since the total mass of pellets a given missile can carry is determined by the altitude the pellets must reach, increasing the number of pellets requires reducing the mass of each pellet to 0.5 g. However, as noted above, it may be possible to shield sensitive parts of the satellite against pellets of this size.

Indeed, to ensure that a pellet penetrates the satellite skin and causes damage if it hits, an attacker may want to use larger pellets. Increasing the pellet mass to 1.5 g (which roughly corresponds to a 1-cm diameter sphere of aluminum) reduces the number of pellets this missile can loft to 600 km to 333,000 and the intercept probability for at least one hit to 20%. If the vulnerable area of the satellite is 15 m^2 rather than 10 m^2, the intercept probability of the base case increases to about 35%.

If there is a systematic error, or bias, in controlling the missile that releases the pellet cloud, this reduces the probability of intercept. Moreover, as noted above, if the launcher is not directly below the path of the satellite, the pellet cloud has a horizontal speed that carries it across the satellite's path, and any a timing errors will cause that motion to result in a spatial inaccuracy that would enter the calculation in the form of a horizontal bias. Unless the value of the bias or the inaccuracy resulting from the timing uncertainty are comparable to the inaccuracy of delivering the cloud, these effects have no significant effect on the intercept probability.

If both the uncertainty in cloud location and satellite location can be reduced to a few tenths of a kilometer, the probability that at least one pellet will hit the satellite is close to one. Notice, however, that both of these uncertainties must be reduced in order to achieve a high intercept probability.

Conclusions

These results show that a simple pellet ASAT of the type considered here, used by an attacker with relatively low technical sophistication, may be relatively ineffective. The effectiveness depends on the values of several key parameters, so that the ASAT would not be very effective unless the attacker can determine the satellite orbit accurately, control the missile accurately, and lift large masses of pellets to orbital altitudes.

Moreover, the outcome of using such a weapon could be very uncertain. The attacker may not even be able to quantify the uncertainty in missile accuracy or predicted satellite location; for example, a country may not have done enough missile tests to determine the missile's accuracy. The attacking country could therefore have little confidence in its ability to carry out a successful attack, even if it fired several missiles at the satellite. As a result, such an ASAT is unlikely to have a high military value.

U.S. planners need to take into account the possibility that such a weapon would work, but their assessment of the threat must consider the attacker's capabilities, as described above. Moreover, the United States could take various steps to make such an attack more difficult. Adding shielding to vulnerable parts of key satellites could defeat such an attack or force the attacker to use pellets with larger mass, which decreases the number of pellets in the cloud for a given payload and reduces the probability of intercept.

Moreover, if the United States detected a missile that appeared to be attacking a satellite, even a relatively small maneuver could essentially eliminate the probability of intercept. The satellite could have more than 300 s to maneuver after the missile was detected, so that a ΔV of only 10 m/s would move the satellite off its predicted orbit by several kilometers, significantly reducing the probability of intercept.[18]

An attacker with a sufficient level of technology could significantly increase the effectiveness of this type of ASAT. If the attacker could both deliver the pellet cloud accurately and accurately determine the satellite's altitude, it could have a high probability of intercept against a target that did not attempt countermeasures, such as maneuvering. A country with these capabilities, however, is likely to have the technical capability to build an interceptor that is more effective and reliable, such as one with a homing interceptor. Given the uncertainties associated with its use, a simple pellet ASAT seems unlikely to be the ASAT of choice.

The Global Positioning System and Reconnaissance Satellites

Section 11 discussed a variety of threats that satellites can face. This subsection puts these threats in perspective by looking in more detail at two important U.S. satellite systems: the Global Positioning System (GPS), which provides global navigation, and the U.S. system used to provide military reconnaissance. As described below, the GPS system has been designed to be robust,

18. This point is also mentioned in the Ensuring America's Space Security (Washington, DC: Federation of American Scientists, 2004), 22, http://www.fas.org/main/content.jsp?form Action=297&contentId=311, accessed December 5, 2005.

resistant to interference, and able to perform its missions even if a few satellites are lost. Moreover, even if the entire satellite system was lost, its capabilities can be provided by other means—at least on a provisional basis. The reconnaissance system is inherently more vulnerable—in part because the satellites are in low earth orbits—but backup capabilities also exist for this mission.

The Global Positioning System

The U.S. military developed GPS as a navigation aid. It remains under military management and performs critical military missions, such as mission planning, guidance of precision munitions, and navigation for troops and vehicles on the ground and in the air. The GPS system has also become integrated deeply into the civil infrastructure. GPS signals are used for civil navigation, for air traffic management, and as a global time standard that synchronizes everything from cell phones to scientific experiments. Degradation or loss of the signal without prior planning could seriously compromise military and economic life.

The GPS constellation consists of 24 operational satellites, with four in each of six different orbital planes at an altitude of 20,000 km.[19] In addition, several spare satellites are usually in orbit, since replacements are launched in advance of the need to replace older ones.[20] Between five and eight satellites are visible from any point on the Earth at all times. However, only four satellites are required to be in view to provide position and time information to a user.[21] If the user can receive signals from more than four satellites, the accuracy improves.

An important feature of the GPS constellation is that its ability to provide navigation information would degrade gradually, rather than catastrophically, under an attack. For example, the minimum

---

19. "USNO GPS Timing Operations," http://tycho.usno.navy.mil/gps.html, accessed January 15, 2005.

20. United States Naval Observatory (USNO), "Block II Satellite Information," ftp://tycho. usno. navy.mil/pub/gps/gpsb2.txt, accessed January 16, 2005. As of January 16, 2005, there were 30 satellites in orbit, with three orbits having 1, 2, and 3 extra satellites, respectively.

21. To obtain the highest accuracy reading with four satellites, they should be positioned so that three satellites are spread evenly around the horizon and one is overhead.

requirement of four satellites would be available for almost the entire day to a user in Beijing even if six satellites were lost, even if those six were chosen to give the greatest loss of service for that location.[22]

The GPS system also has some innate protections from attack. The satellites operate at a high altitude, which puts them out of reach of ground-based kinetic energy attacks using modified short-range or intercontinental ballistic missiles. At this altitude, the space environment is difficult to operate in: the flux of charged particles from the Van Allen belts is high (see Section 5), and any long-lived ASAT that was intended to trail a GPS satellite would need to be designed to handle this danger. Blinding and dazzling would not affect the ability of the satellites to provide navigation information since doing so requires no optical sensors.[23] The satellites themselves are designed to withstand heat imbalances from distant laser attacks.[24] Moreover, the satellites are in six separate orbital planes and their spacing within those orbits is such that a single ASAT would have difficulty targeting more than one satellite.

Uplink jamming of the command signals from ground stations is difficult. However, even if an attacker succeeded in jamming the uplink, the satellites are able to operate for 14 days without contact from the command station and up to 180 days in an autonomous navigation mode (AUTONAV). Moreover, the most recent version (Block IIR) can do so while maintaining full accuracy.[25] In autonomous mode, the satellites communicate only with each other using cross links, which are protected against jamming by using frequency hopping and directional antennas that can receive signals from other satellites but not the ground.

22. Geoffrey Forden, "Appendix D: Sensitivity of GPS Coverage to Loss of One or More Satellites," Ensuring America's Space Security (Washington, D.C.: Federation of American Scientists, 2004), http://www.fas.org/main/content.jsp?formAction=297&contentId=311, accessed January 15, 2005. Forden finds that four satellites would be observable from Beijing for 22 hours of the day.

23. Each GPS satellite carries optical sensors as part of the Nuclear Detonation Detection system, but these are not used for navigation.

24. Ashton B. Carter, "Satellites and Anti-Satellites: The Limits of the Possible," International Security, 10 (1986) 90.

25. USNO, "Block II Satellite Information."

Three types of interference attacks that might succeed are downlink jamming, and to a lesser degree, spoofing and meaconing.

Downlink jamming is not technically demanding and can lead to significant interference locally. Sites on the internet provide blueprints for a GPS jammer that could be built by someone with an undergraduate technical degree,[26] as well as advertisements for GPS jammers for sale. A small, lightweight, short-lived jammer that could deliver up to 100 W of emitted power can be purchased or constructed for less than $1,000. Jammers that can deliver kilowatts of power can be made for about $100,000, and those that produce tens to hundreds of kilowatts would cost a million dollars or more.[27] However, such high power jammers are likely to be vulnerable, as they can be located rapidly and accurately and then targeted.

Multiple smaller jammers would likely be the preferred scheme. They could be readily distributed in large numbers over terrain in which an enemy wished to deny GPS tracking. If a jammer can send its signal over a wide area without obstruction—for example, by broadcasting from an airplane—it could affect a large number of users. An airborne jammer emitting only 1 W of power could deny GPS tracking to active users (those who have already locked onto the GPS signal) within 10 km and could deny users up to 85 km away the ability to locate GPS satellites and receive an accurate signal.[28] In addition, before a GPS receiver can recognize that it is being jammed and cease providing position data, it may provide inaccurate data for tens of seconds, thereby endangering pilots who do not employ backup navigation measures.

The United States has taken some steps to mitigate the GPS system's vulnerability to jamming. The GPS modernization program will increase the strength of the signal broadcast from

26. Marks ("Wanna Jam It?") describes an Air Force team that built a jammer to work against an ultrahigh frequency satellite, with just an internet connection and $7,500 worth of materials. It would be fairly simple to adapt the same technique to the GPS frequency.

27. John A. Volpe National Transportation Systems Center, "Vulnerability Assessment of the Transportation Infrastructure Relying on the Global Positioning System," Appendix A, August 29, 2001, and references therein.

28. P. Ward, "GPS Receiver RF Interference Monitoring, Mitigation, and Analysis Techniques," Navigation—Journal of the Institute of Navigation, 414 (1994): 367.

the satellites and the number of frequencies the signal is broadcast on. This will raise the bar for jammers, requiring additional signal strength and the ability to jam multiple frequencies. New GPS receivers are being built with additional antijamming features. For example, by using multiple antennas, an analog receiver can use nulling techniques to eliminate the interfering signal, and new digital receivers can be made even more jam resistant.[29]

Spoofing, as Section 11 indicates, mimics the characteristics of a true signal so that the user receives the fake signal instead of the real one. Meaconing is similar to spoofing but entails receiving the real signal and then rebroadcasting it with a time delay. Because two GPS signals are broadcast at different frequencies, a civilian signal and a military signal that is encrypted, it is possible to interfere with one but not the other.

The potential advantage of spoofing and meaconing over jamming is that they could be done covertly, since the user will generally not be able to determine that the receiver is being spoofed or meaconed. In both cases, the antenna would be more difficult to locate than a jammer's antenna, since the signal strength of the spoofed or meaconed signal must be the same as that of the true GPS signal. If the location of the spoofing or meaconing antennas were determined, receivers could be told to ignore the false signals.

Spoofing a GPS satellite signal is technically much more difficult than jamming it, as the spoofed signal must have the same characteristics as the real one, including frequency structure and signal encoding. Because it is encrypted, the GPS military signal would be extremely difficult to spoof, but the civil signal has a well-known structure. Moreover, GPS satellite simulators, which are used by manufacturers to test GPS receivers and related products, are available commercially and can be purchased (for $10,000 to $50,000) or rented. These simulators produce fake satellite signals at a higher power than the real GPS civil signals. Someone with an advanced understanding of electronics could also build a GPS simulator from scratch using information available on

29. Robert K. Ackerman, "Jam-Proof Signals to Guide Navigation," Signal Magazine, November 2001, http://www.afcea.org/signal/articles/anmviewer.asp?a=481&z=6, accessed February 9, 2005.

the internet.[30] There are some technical countermeasure strategies that the GPS receiver could implement relatively easily, but none are in wide use.[31] Meaconing also requires considerably more technical sophistication than jamming, but may be easier to use against the encrypted military systems.

Jamming could be made even more difficult in a theater conflict by using pseudosatellites or pseudolites, which are high power GPS transmitters that can be deployed on ground systems or unmanned aerial vehicles (UAVs).[32] These transmitters would broadcast a signal locally that would be a great deal more powerful than the satellite signals.

If the GPS civil signal is interfered with in one of these ways, users may have other options for navigation aid and precision guidance. The European Union is developing an independent satellite navigation system, Galileo, which is similar to GPS. The Galileo system was originally planned to have an operating frequency band that would overlap the U.S. encrypted military signal. This would have prevented the United States from jamming the Galileo signal without also jamming its own military signal. This was an intensely negotiated aspect of the Galileo system, and it appears that a compromise has been reached in which the Galileo system will not interfere with U.S. military operations.[33]

The Effects of a Diminished GPS Capability.[34] As GPS is so widely used in military and civilian activities, an exhaustive discussion of what effects temporary outages and diminished

---

30. Roger G. Johnston and Jon S. Warner, "Think GPS Cargo Tracking = High Security? Think Again," September 24, 2003, http://www.eyefortransport.com/ index. asp?news=38732& nli= freight&ch, accessed January 16, 2005.

31. Jon S. Warner, Roger G. Johnston, "GPS Spoofing Countermeasures," December 2003, at http://www.homelandsecurity.org/bulletin/ Dual%20 Benefit/warner_gps_ spoofing. html, accessed January 8, 2005.

32. Maryann Lawlor, "Researchers Locate Satellite Options," Signal Magazine, November 2001, http://www.afcea.org/signal/articles/anmviewer.asp?a=480, accessed February 9, 2005.

33. Robert Wall and Michael A. Taverna, "Navigating Hurdles: U.S. and Europe ink agreement on coexistence of GPS and Galileo," Aviation Week & Space Technology, June 28, 2004, 31.

34. This analysis is based on "Vulnerability Assessment of the Transportation Infrastructure."

accuracy of GPS would have is beyond the purpose of this report. Instead, impacts on a few GPS applications are examined here.

Temporary outages of GPS may not have a serious impact on ground- and sea-based navigation, especially if the outages were detected and backup systems put into use smoothly. The temporary loss of GPS may be least problematic for ships at sea, since they move slowly. Exceptions include situations where this loss is combined with another aggravating effect, such as low visibility, bad weather, or complicated terrain.

For air navigation, such outages would be more serious, especially in situations with high traffic or difficult terrain. Frequent and random outages from, for example, small and numerous jammers may make traffic control chaotic and dangerous. This could be particularly dangerous for civilian air traffic, which would not be as accustomed to functioning in crisis situations as is military traffic. However, for this reason, such vulnerable systems do not now and are unlikely in the future to rely solely on GPS systems.

If a threat to GPS seemed imminent, civilian applications such as surface shipping and business coordination can implement backups. Systems currently extant in the United States should suffice for this purpose. For example, LORAN-C[35], a ground-based system that transmits navigation signals, can provide two-dimensional navigation in the continental United States and serve as an accurate time standard. However, it cannot replace systems that require high precision in three dimensions, including some aspects of aviation. Although GPS was scheduled to replace LORAN-C, a December 8, 2004, presidential directive makes it more likely that it will be retained as a backup system. This directive gave the Department of Homeland Security responsibility for developing contingency responses in case GPS is disrupted within the United States.[36]

35. See the U.S. Coast Guard Navigation Center's web pages on LORAN-C beginning at http://www.navcen.uscg.gov/loran/default.htm, accessed December 15, 2004.

36. "New Presidential GPS Policy Elevates Executive Oversight, Security Issues," GPS World, January 1, 2005, http://www.gpsworld.com/gpsworld/article/articleDetail.jsp?id=140849 accessed on January 16, 2005.

Reconnaissance Satellites

Satellite reconnaissance is used to perform numerous strategic and tactical military missions, including mapping terrain, gathering information on the military and industrial capabilities of other countries, monitoring one's own troop movements, choosing targets during a conflict, and assessing battle damage.

The United States has a number of dedicated military reconnaissance satellites: three optical imaging reconnaissance satellites, with ground resolution reported to be 12–15 cm; and three synthetic aperture radar satellites, with ground resolution reported to be roughly a meter.[37] (There are a number of signals intelligence satellites, too—satellites that detect radio signals.) Many other countries operate reconnaissance satellites as well.

Ownership of reconnaissance satellites is not restricted to governments. A few commercial satellites take optical and infrared images useful for intelligence work. For example, the French SPOT system of satellites[38] takes images of the ground with up to 2.5-m resolution. The EROS-A[39] satellite can deliver images from 1- to 1.8-m resolution. The Ikonos satellite[40] provides images with up to 1-m resolution. The U.S.-based Quickbird satellite[41] provides images with resolution below 1 m. In principle, these satellites provide imaging data to anyone who will pay for them. In practice, a country could exercise "shutter control," as the United States did during the beginning of the war in Afghanistan by buying the exclusive rights to the images in some parts of the Ikonos orbit.

---

37. Robert Windrem, "Spy Satellites Enter New Dimension," MSNBC News, October 9, 2001, http://msnbc.msn.com/id/3077885/, accessed January 31, 2005.

38. SPOT 4 and SPOT 5 are currently in orbit. See Spot Image, http://www.spotimage.fr, accessed January 18, 2005.

39. EROS-A, launched on December 5, 2000, is owned ImageSat International N.V., a commercial endeavor founded by a team of engineers from the Israeli space program. See ImageSat International, http://www.imagesatintl.com/index.shtml, accessed January 18, 2005.

40. Space Imaging provides optical images obtained with the Ikonos satellite; see Space Imaging, http://www.spaceimaging.com, accessed January 18, 2005.

41. Digitalglobe provides images obtained with the Quickbird satellite; see DigitalGlobe, http://www.digitalglobe.com/, accessed January 18, 2005.

Remote sensing satellites are vulnerable to blinding and dazzling attacks from ground-based lasers, as Section 11 discussed. Because they are usually in low earth orbits, they are also vulnerable to kinetic energy attacks launched by ballistic missiles.

If remote sensing satellites are compromised, some of their functions can be provided by other systems, especially for regional or tactical use. For example, unmanned aerial vehicles (UAVs) can be used to augment satellite reconnaissance. In a conflict where the area of interest is confined to a theater of operations, the reduced field of view available at these lower altitudes (see Figure 5.2) may be compensated for by using multiple UAVs.[42] An airplane-based radar system, such as the U.S. Air Force synthetic aperture radar system JSTARS,[43] could also be used for tactical reconnaissance. Using UAVs and airplane- based radars is viable only in a region where the party has air superiority.

It may also be possible to use satellite-based systems to provide backup tactical reconnaissance capabilities. For example, the United States is developing launch vehicles that could launch small payloads with a minimum of preparation time. Such launch vehicles could put into orbit imaging satellites that are smaller and less expensive than current reconnaissance satellites. These satellites could be in lower orbits than the current reconnaissance satellites to compensate for their lower power optics and thus provide adequate ground resolution.

42. See Bruce M. DeBlois et al., "Space Weapons: Crossing the U.S. Rubicon," International Security 29 (fall 2004): 61.

43. For more information, see the U.S. Air Force Fact Sheet on E-8C Joint Stars, http://www. af.mil/factsheets/factsheet.asp?fsID=100, accessed January 31, 2005, and the Air Force Technology.com's web page on JSTARS, http://www. airforce-technology.com/projects/jstars/, accessed January 31, 2005.

Appendix - 12

Calculating the Effectiveness of a Pellet ASAT

This section describes how to calculate the probability that at least one pellet strikes the satellite. The parameters that enter this calculation are

- the uncertainty in the location of the pellet cloud
- the uncertainty in the location of the satellite
- the number of pellets in the cloud
- the size of the pellet cloud
- an area that represents the vulnerable frontal area of the satellite.

The cloud radius is varied for fixed values of the other parameters, giving the intercept probability as a function of cloud radius, R_c. The maximum of this function gives the intercept probability at the optimum cloud radius. The pellet cloud is assumed to be spherical. While the shape of the pellet cloud could be optimized to increase the probability of intercept, doing so would require that the country be able to control the orientation of the cloud; otherwise, doing so could instead decrease the intercept probability. Since the case of interest is that of a low-tech ASAT, the attacker is assumed to use a spherical cloud.

The problem can be analyzed as a two-dimensional problem in a plane lying perpendicular to the satellite's trajectory. The spherical pellet cloud is then replaced by a disk of radius R_c with an area density of pellets equal to the two-dimensional projection of spherical density. For simplicity, in this calculation this two-dimensional projection of the density is replaced with a uniform density across the disk.[1]

The satellite is assumed to pass directly over the ASAT launch site; in this case, arranging the pellet cloud to have its apogee at the altitude of the satellite's trajectory eliminates timing errors. If the satellite does not pass over the ASAT launch site, inaccuracies in cloud position due to timing errors are added to the calculation as an increased uncertainty in cloud position. It is useful to describe the calculation in two steps. The first step assumes the attacker knows exactly the location of the satellite and illustrates the effect of uncertainty in the location of the pellet cloud. The second step includes the uncertainty in satellite location.

If the attacker knows the position of the satellite at the predicted intercept, the ASAT is aimed at that point. Because of errors in guiding the missile perfectly, the center of the pellet cloud instead arrives some distance from the aim point. The uncertainty in the location of the cloud is described by a Gaussian distribution:

$$p(d) = \frac{1}{2\pi\sigma_c^2} e^{-d^{2}/2\sigma_c^2} \qquad (12.1)$$

where d is the distance between the center of the pellet cloud and the aim point in a plane perpendicular to the satellite's trajectory, and σ_c describes the uncertainty in the location of the pellet cloud due to the inaccuracy of the missile launching it, assuming the bias is zero (see Figure 12.2). The center of the cloud lies within a circle of radius σ_c roughly 40% of the time.

---

1 Consider the results of dispersing the same number of particles uniformly in a disk and a sphere of the same radius. Let ρ be the area density of particles in the disk. Compare this to the effective area density that would result from projecting the spherical density onto a plane perpendicular to the direction of motion of the satellite; this is the physical quantity that matters as the satellite sweeps through the cloud. The effective area density is 1.5ρ at the center of the sphere and falls to zero at the edge of the sphere; it is 1.3ρ halfway to the edge of the sphere, and 0.99ρ three-quarters of the way to the edge.

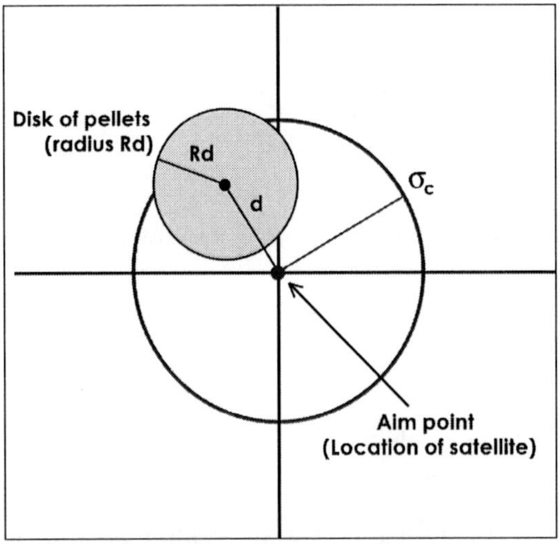

Figure 12.2. This figure shows a plane perpendicular to the path of the satellite; the satellite passes through the plane at the point marked "aim point." While the ASAT is aimed at this point, inaccuracies cause the center of the cloud to arrive at a distance d from the aim point. Roughly 40% of the time, the center of the cloud falls within the circle marked σ_c.

The probability that a pellet hits the satellite is the product of the probability that the satellite passes through the cloud, multiplied by the probability that a pellet hits the satellite if the satellite passes through the cloud. The first probability is just the probability that d is less than or equal to the cloud radius R_c, which is given by integrating $p(d)$ in Equation 12.1 for d between 0 and R_c; this integration gives

$$P_{satellite\,in\,cloud} = 1 - e^{-R_c^2/2\sigma_c^2} \qquad (12.2)$$

Next we calculate the probability that the satellite is hit by at least one pellet if it passes through the cloud. As discussed above, for simplicity assume that the area density of pellets in the plane perpendicular to the path of the satellite is constant throughout the cloud. Let N be the total number of particles in the cloud and A_S be the area of the satellite perpendicular to its direction of motion. Each pellet has equal probability of being anywhere in the disk of area $R_c^{\ 2}$. As a result, the probability that n pellets hit the satellite, assuming the satellite passes through the pellet cloud, is given by the binomial distribution

$$P_{n\,hits} = \frac{N!}{n!(N-n)!}p^n(1-p)^{N-n},\ p = \frac{A_s}{\pi R_c^2} \qquad (12.3)$$

For large N, this approaches the Poisson distribution[2]

$$P_\lambda(n) = \frac{\lambda^n e^{-\lambda}}{n!},\ \lambda = Np = \frac{NA_s}{\pi R_e^2} \qquad (12.4)$$

The probability that the satellite is hit by at least one pellet if it passes through the pellet cloud is

$$P_{\geq 1\,hit} = 1 - P_\lambda(0) = 1 - e^{-\lambda} \qquad (12.5)$$

Similarly, the probably that the satellite is hit by at least two pellets if it passes through the pellet cloud is

$$P_{\geq 2\,hit} = 1 - P_\lambda(0) - P_\lambda(1) = 1 - (1+\lambda)\,e^{-\lambda} \qquad (12.6)$$

The total probability that at least one pellet hits the satellite is then given by the product of Equations 12.2 and 12.5.

Step 2

In reality, the attacker does not know the exact location of the satellite, but can determine it only with some uncertainty. The aim point of the ASAT is therefore the best estimate of the satellite's position. In the plane perpendicular to the path of the satellite, the uncertainty around this best estimate is described by a two-dimensional Gaussian probability:

$$r(d_x, d_z) = \frac{1}{2\pi\sigma_x\sigma_z} e^{-\frac{1}{2}\left[\left(\frac{d_x}{\sigma_x}\right)^2 + \left(\frac{d_z}{\sigma_z}\right)^2\right]} \qquad (12.7)$$

Here $r(d_x, d_z)$ is the probability that the satellite is actually located a distance d_x and d_z from the best estimate in the horizontal and vertical directions, respectively, where σ_x and σ_z describe the widths of the Gaussian uncertainty in these two directions. In general, these widths are not the same: the uncertainty in altitude

2 "Poisson Distribution," http://mathworld.wolfram.com/PoissonDistribution.html, accessed February 10, 2005.

(σ_z) is generally several times larger than the uncertainty in horizontal position *(σx)*, for the assumed case of optical tracking.

Calculating the probability of a pellet hitting the satellite including both the uncertainty in the location of the pellet cloud and of the satellite involves the following steps. To determine the probability that the satellite passes through the debris cloud, we first calculate the probability that the actual location of the satellite is such that it would pass through a pellet cloud located at a particular point (x_c, z_c). We then integrate over all possible locations of the pellet cloud. Finally, we multiply by the probability that at least one pellet hits the satellite if the satellite does pass through the cloud (see Figure 12.3).

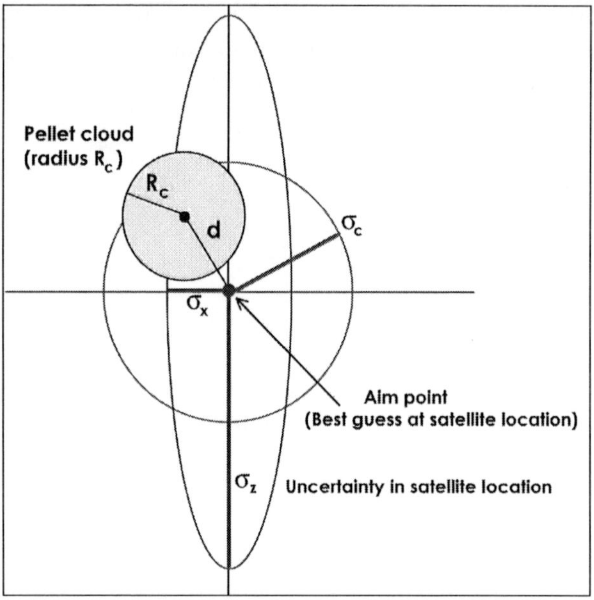

Figure 12.3. This figure adds to Figure 12.2 an ellipse giving the uncertainty in the location of the satellite's position.

More specifically, we assume that the missile delivers the pellet cloud so that its center is at the point (x_c, z_c), and that the cloud has a radius R_c at the time the satellite crosses the plane containing the center of the cloud. The probability that the satellite passes through a cloud at this location is found by integrating the function *r(x,z)* describing the location of the satellite (Equation

12.7) over the area of the pellet cloud, i.e., over a disc of radius R_c centered at (x_c, z_c). This integral, which is a function of x_c and z_c, is then multiplied by the function from Equation 12.1, which gives the probability that the center of the cloud is located at (x_c, z_c), then the product is integrated over the entire plane, *i.e.*, over all possible cloud locations. This integration gives the total probability that the satellite passes through the cloud.

Multiplying this result by the probability that the satellite will be struck by at least one pellet if it passes through the cloud (Equation 12.5) gives the total probability that the satellite is struck by at least one pellet, when the cloud radius is R_c. Multiplying by Equation 12.6 gives the probability of at least 2 hits. The calculation is repeated for a range of values of R_c, and these points are plotted to find the optimum cloud size and the corresponding maximum intercept probability.

Contributors

Laura Grego is a Staff Scientist in the Global Security Program at the Union of Concerned Scientists (UCS). Before joining UCS, she was a researcher at the Harvard- Smithsonian Center for Astrophysics. She received her Ph.D. in physics from the California Institute of Technology in 1999.

Lisbeth Gronlund is Co-Director and Senior Scientist in the Global Security Program at UCS and a Research Scientist in the Security Studies Program at the Massachusetts Institute of Technology (MIT). Previously, she was an SSRC-MacArthur Foundation Fellow in International Peace and Security at the Center for International Security Studies at the University of Maryland and a postdoctoral fellow at the MIT Defense and Arms Control Studies Program. She is a fellow of the American Physical Society and was a co-recipient of their Joseph A. Burton Forum Award in 2001. Gronlund received her Ph.D. in physics from Cornell University in 1989.

David Wright is Co-Director and Senior Scientist in the Global Security Program at UCS and a Research Scientist in the Security Studies Program at MIT. Prior to coming to UCS, he was an SSRC-MacArthur Foundation Fellow in International Peace and Security at the Center for Science and International Affairs in the Kennedy School of Government at Harvard, and a Senior Analyst at the Federation of American Scientists. He is a fellow of the American Physical Society and was co-recipient of their Joseph A. Burton Forum Award in 2001. Wright received his Ph.D. in physics from Cornell University in 1983, and worked as a research physicist for five years before beginning full-time work on security issues.

Index

Other Valuable Books of Related Interest

Conspiracy to Kill Rajiv Gandhi: From CBI Files, by K Ragothaman (Former SP - CBI)

India's Spy Agencies: Shaken not Stirred, by Sunil S Parihar

Amazing Ladakh: Travel, trckking, religion, culture, wildlife almost everything about Leh-Kargil-Zanskar-Siachen, by Parvez Dewan, IAS

Heroes of North East India, by Sumitra Mukherjee

The 21st Century Army: Strategies for Future, by Lt Gen Gautam Banerjee, PVSM, AVSM, YSM

Jungle Warfare: An Ultimate Handbook by Dr NC Asthana, IPS

Ultimate Handbook of Urban Warfare by Dr NC Asthana, IPS

Army Leadership, by Department of Army, US

Counter Insurgency Guide, by Bureau of Political-Military Affairs, US

Lens and the Guerrilla: Insurgency in India's Northeast, by Rajeev Bhattacharya **(fully Illustrated)**

Intelligence Tradecraft: Secrets of Spy Warfare, by MK Dhar (former Joint Director, IB)

Manas Police & Security Yearbook 2010-2011 (fully Illustrated) by Ajit Doval IPS, BR Lall, IPS and other Top Cops.

Manas Defence Year Book 2010-2011 (Fully Illustrated) by General S Padmanabhan and Lt Gen S Pattabhiraman

CBI Top-Cop Recalls by SK Dutta, IPS (Former Director – CBI)

Pakistan's Intelligence by Sudhakar Raje (Ex-Indian Diplomat)

Who Runs Police? Secrets of a Top Cop by HS Swan, IPS (Ex-DGP, Haryana)

Is It Police? Confessions of a Top Cop by Vinay Kumar Singh, IPS

Police and Naxalism: Strategies for Solutions by ML Kumawat, IPS, Nazrul Islam, IPS and others

Common Man's Guide to Police & Criminal Laws by Dr DV Guruprasad, IPS (DG – Karnataka Police)

Rainbow People: Reinventing Northeast India by Falguni Rajkumar, IAS (Former Sec NEC)

Industrial Security: Management & Strategies by DC Nath, IPS & Capt SB Tyagi

Intelligence, Security & Asymmetric Warfare by Dhirendra Singh, IAS and Lt Gen DB Shekatkar

Intelligence for Management by DC Pathak, IPS (Former Dir-Intelligence Bureau)

Inside IB and RAW: The Rolling Stone that Gathered Moss by K Sankaran Nair, IPS (Former Head: Research and Analysis Wing (RAW)

Intelligence: Corporate Success and Vigilance by D C Pathak, IPS (Ex Chief – IB)

The Undisclosed Punjab: India Besieged by Terror by Birbal Nath, IPS (Former DGP, Punjab)

The Folds of Insurgency by AP Maheshwari, IPS

Who Owns CBI? The Naked Truth by BR Lall, IPS (Former: Joint Director, CBI)

Saga of Civil Services by Dr Yogendra Narain, IAS (Former Defence Secretary)

Open Secrets: India's Intelligence Unveiled by M K Dhar, IPS

Security Manual by P Benjamin

SPYMASTER: My Life in the CIA by Ted Shackley

VIP Security: What You Need to know by Kanchan Sinha

Armed Forces Tribunal by Dr UC Jha

Beyond North East: Insurgency-History-Development by HN Das (Former Chief Secretary, Assam)

Why Bureaucracy Fails by Kalyan Ray, IAS

WMD Terrorism: Strategy for Consequences Management by Brig MS Patial

Intelligence Manual by Department of Defence, USA

The ISI Game Plan by Brig Yash Mohan, VSM

Terror, Insurgency and the State: Ending Protracted Conflicts by Marianne Heiberg etc.

Operation Triple X: An Indian Spy-Run in Pakistan A Novel by MK Dhar, IPS

Naxalism: Causes and Cure by PK Agrawal, IAS

Legal Safeguards for Defence Personnel by Air Cmde. Ranvir Kumar (Former Judge Advocate General)

Reforming Intelligence Agencies: Pakistan's Transitional Democracy by Frederic Grare

Combating Terrorism: Strategies of Ten Countries by Yonah Alexander

Strategies in Defence and Intelligence by Lt Gen S Pattabhiraman & MK Dhar, IPS

Sports, Games and Adventure Sports by Air Cmde (Dr) CN Ghosh

The Craft of Intelligence by Allen W Dulles

Fulcrum of Evil: ISI-CIA-Al Qaeda Nexus by MK Dhar (Former Joint Chief of IB)

The Conspirators: Fiction of Facts (A Book on Pakistan's Intelligence) by Brig Yash Mohan, VSM

Silent Warfare: Understanding the World of Intelligence by Abram N Shulsky & Gary J Schmitt

Wages of Crime: Black Markets, Illegal Finance and the Underworld Economy by R T Naylor

The Peace Keepers: Indian Police Service (IPS) by SR Arun, IPS

Para-Military Forces: Central Reserve and Foreign Policy by Dr AK Pasha

Human Rights Training by Dr S Subramanian, IPS

Industrial Espionage: Causes and Cure by Dr S Subramaniam, IPS

Executive Protection by Dr S Subramaniam, IPS

CIA's Eye on South Asia by Anuj Dhar, Journalist

Threat to Security: How Secured is India from within by PC Dogra, IPS (Former DGP, Punjab)

Effective Interrogation: Naxalism, Police, Intelligence by G Chatterjee (Former Intelligence Officer)

Financial Terrorism: Black Money and the Indian Elite by BR Lall, IPS (Former: Joint Director, CBI)

White Collar Crimes X-Posed by Thakur Shailendra Nath